ARCO

Everything you need to score high on

SAT II Math

8th Edition

ARCO

Everything you need to score high on

SAT* II Math

8th Edition

MORRIS BRAMSON, M.S.
and
NORMAN LEVY, Ph.D.

MACMILLAN • USA

* SAT is a registered trademark of the College Entrance Examination Board, which does not endorse this book.

Eighth Edition

Macmillan General Reference
A Simon & Schuster Macmillan Company
1633 Broadway
New York, NY 10019-6785

Macmillan Publishing books may be purchased for business or sales promotional use. For information please write: Special Markets Department, Macmillan Publishing USA, 1633 Broadway, New York, NY 10019-6785.

Copyright © 1998, 1995, 1991, 1987, 1982, 1977, 1976, 1966, by ARCO Publishing,
a division of Simon & Schuster, Inc.
All rights reserved
including the right of reproduction
in whole or in part in any form.

An ARCO Book

MACMILLAN is a registered trademark of Macmillan, Inc.
ARCO is a registered trademark of Prentice-Hall, Inc.

Library of Congress Number: 97-81132

ISBN: 0-02-862473-4

Manufactured in the United States of America

10 9 8 7 6 5 4 3 2 1

CONTENTS

PART 1
What You Should Know About SAT II: Subject Tests

The Importance of Subject Tests for College Admission 3
High Marks in School Are Not Enough 3
The SAT Program 3
What the Mathematics Tests Are Like 4
How to Prepare for Your Math Test 4
How to Take the Test 5
What Your Mathematics Test Score Means 5
Applying for the Examination 6
Rules of Conduct 6
Transmitting the Results 7

PART 2
Outline of Topics

I. Arithmetic 11
II. Algebra 11
III. Geometry 12
IV. Functions and Their Graphs 13
V. Real Number System 14
VI. Logic 14
VII. Sets 15
VIII. Trigonometry 15
IX. Miscellaneous Topics 16

PART 3
Math Review—Illustrative Problems and Solutions

1. Formulas and Linear Equations 19
2. Algebraic Fractions 21
3. Sets 24
4. Functions 25

5. Exponents 28
　　6. Logarithms 31
　　7. Equations—Quadratic, Radical and Exponential 33
　　8. Inequalities 37
　　9. Verbal Problems 39
　10. Geometry 42
　11. Trigonometry 56
　12. Graphs and Coordinate Geometry 64
　13. Number Systems and Concepts 67
　14. Arithmetic and Geometric Progressions 71
　15. Vectors 74
　16. Variation 76

PART 4
Math Practice Exercises and Solutions by Topic

　　1. Formulas and Linear Equations 81
　　2. Algebraic Fractions 81
　　3. Sets 82
　　4. Functions 83
　　5. Exponents 83
　　6. Logarithms 84
　　7. Equations—Quadratic and Radical 85
　　8. Inequalities 85
　　9. Verbal Problems 86
　10. Geometry 86
　11. Trigonometry 88
　12. Graphs and Coordinate Geometry 89
　13. Number Systems and Concepts 91
　14. Arithmetic and Geometric Progressions 92
　15. Vectors 92
　16. Variation 93
Solutions to Practice Exercises 94

PART 5
Four Sample Mathematics Tests: Level IC

Steps to Take After Each Sample Test 127
Sample Test 1: Math Level IC 131
　Answer Key 141
　Solutions 141

Sample Test 2: Math Level IC 155
- Answer Key 164
- Solutions 164

Sample Test 3: Math Level IC 179
- Answer Key 188
- Solutions 188

Sample Test 4: Math Level IC 199
- Answer Key 208
- Solutions 208

PART 6
Three Sample Mathematics Tests: Level IIC

Sample Test 1: Math Level IIC 221
- Answer Key 228
- Solutions 228

Sample Test 2: Math Level IIC 243
- Answer Key 251
- Solutions 251

Sample Test 3: Math Level IIC 263
- Answer Key 271
- Solutions 271

PART ONE

What You Should Know About SAT II: Subject Tests

CONTENTS

The Importance of Subject Tests for College Admission   3
High Marks in School Are Not Enough   3
The Nature of College Entrance Tests   3
What the Mathematics Tests Are Like   4
How to Prepare for Your Math Test   4
How to Take the Test   5
What Your Mathematics Test Score Means   5
Applying for the Examination   6
Rules of Conduct   6
Transmitting the Results   7

The Importance of Subject Tests for College Admission

Many of our nation's colleges insist that applicants take one or more SAT II: Subject Tests. If you are applying to a school which requires you to take Subject Tests, you should be aware that the results of the tests are not the sole factor in determining if you will be admitted. Other factors come into play: your SAT I scores, high school scholastic record, standing in your graduating class, grades in specific high school subjects, and the personal interview. Doing well on your Subject Tests, however, may substantially increase your chances of being accepted by the college of your choice.

The Subject Tests are administered throughout the world, and thousands take the exams annually. The College Entrance Examination Board (CEEB), which administers the tests, will send to the college admissions officer not only your score on the test you take, but also your percentile ranking. The latter tells how many test-takers did better than you and how many did worse. It follows, therefore, that the admissions officer seriously considers your standing on each Subject Test that you offer to determine how well you are likely to do in college work.

High Marks in School Are Not Enough

Since secondary schools have varying standards of grading, it is understandable that high school marks alone will not suffice when colleges try to appraise objectively the ability of an undergraduate to do college work. An "A" in a course of English in High School X may be worth a "C" in High School Y. Moreover, teachers within the same high school differ among themselves in grading techniques. The Subject Tests are highly objective. Consequently, they have become a *sine qua non* for many college admissions officers in order to predict success or lack of success for applicants.

The SAT Program

The SAT program consists of the following parts:

1. SAT I: Reasoning Test

2. SAT II: Subject Tests

SAT I

The SAT I provides a measure of general scholastic ability. It is not an intelligence test nor is it, in the strict sense, an achievement test. It yields two scores: verbal ability and mathematics ability. Included in the test are verbal reasoning questions, reading comprehension questions drawn from several fields, and various kinds of quantitative-mathematical materials. These include questions on arithmetic reasoning, on algebraic problems, and on the interpretation of graphs, diagrams, and descriptive data. The SAT I takes three hours to answer questions plus the time to collect and check testbooks and to allow for a rest period.

SAT II: SUBJECT TESTS

Subject Tests are given in the following subjects:

American History and Social Studies	Korean
Biology	Latin
Biology E/M	Literature
Chemistry	Mathematics Level IC
Chinese	Mathematics Level IIC
English Proficiency	Modern Hebrew
French	Physics
German	Spanish
Italian	World History
Japanese	Writing

What the Mathematics Tests Are Like

Each Mathematics Subject Test consists of 50 multiple-choice questions with answer choices from (A) to (E). The questions are designed to test the mathematical competence of students who have studied college-preparatory mathematics—Level I for three years of study and Level II for more than three years of study. Obviously, an examination testing three or more years of study touches very briefly on a great many topics and concepts. Both levels of the Mathematics Subject Tests measure understanding of elementary algebra, three-dimensional geometry, coordinate geometry, statistics, and basic trigonometry. The Level II exam tests these topics with more advanced content. For example: While the Level I exam includes questions on plane geometry, the Level II exam covers transformations and coordinate geometry in two or three dimensions; while the Level I trigonometry questions are based on right-triangle trigonometry and the fundamental relationships among the trigonometric ratios, the Level II test examines understanding of the properties and graphs of the trigonometric functions, the inverse trigonometric functions, trigonometric equations and identities, and the laws of sines and cosines; Level I functions are mainly algebraic functions, while Level II functions extend to the more advanced logarithmic and exponential functions; Level II statistics go beyond mean, median, mode, counting, and data interpretation to questions on probability, permutations, and combinations; the miscellaneous topics covered at Level II go beyond simple logic, elementary number theory, and arithmetic and geometric sequences to include logic and proofs and limits.

In addition, the emphasis on various topics varies between the two levels. Half of the questions at Level I are directed at algebra and plane geometry and another quarter of the questions measure understanding of coordinate geometry and functions. At Level II, on the other hand, plane geometry is not tested at all, but nearly half of the questions are concentrated on trigonometry and functions. Level II devotes twice as many questions to miscellaneous topics as does Level I.

How to Prepare for Your Math Test

Let us sound a clear warning: *Don't wait till a week or even a month before the examination to start your preparation for it.* Cramming is not recommended. The best preparation is intensive review over a period of several months.

Familiarity with the types of questions on this test will unquestionably prove helpful. For this reason, we advise you to use this book in the following way:

First, carefully read Part Two, *Outline of Topics*. This chapter gives you an illuminating cross-section of the mathematics areas which you will find on your test.

After you have read this part, choose your level and take your first sample test. The sample tests in this book are carefully patterned on the actual Mathematics Subject Tests. They are designed to familiarize you with the types and difficulty level of questions which you will face on the actual examination.

Put yourself under strict examination conditions, and allow yourself exactly one hour of working time for each sample exam.

Tolerate no interruptions while you are taking a Sample Test. Work in a steady manner. Do not spend too much time on any one question. If a question seems too difficult, proceed to the next one. If time permits, go back to the omitted question.

Do not place too much emphasis on speed. The time element is a factor, but is not all-important. Accuracy should not be sacrificed for speed.

Use the answer key provided at the end of each sample test to score yourself following the instructions given in the chapter "Steps To Take After Each Sample Test." Identify the nature of each question that you answered incorrectly or omitted and look up the related topic in Part Three, *Mathematics Review—Illustrative Problems and Solutions*. Study the review material and *Illustrative Problems* related to this topic and check the solutions provided. Consult with teachers or textbooks as needed. Then proceed to another sample test.

How to Take the Test

Do not become disturbed if you find yourself unable to answer a number of questions in a test, or if you are unable to finish. No one is expected to achieve a perfect score. There are no established "passing" or "failing" grades. Your score compares your performance with that of other candidates taking the test, and the report to the college shows the relation of your score to theirs.

Although the test stresses accuracy more than speed, it is important that you use your time as economically as possible. Work as steadily and rapidly as you can without becoming careless. Take the questions in order, but do not waste time pondering questions which for you contain extremely difficult or unfamiliar material.

Read the directions with care. If you read too hastily, you may miss an important direction and thus lose credit for an entire section.

SHOULD YOU GUESS ON THE TEST?

A percentage of wrong answers is subtracted from the number of right answers as a correction for haphazard guessing. Mere guessing will not improve your score significantly, and may even lower it. If you are not sure of the correct answer but have some knowledge of the question and are able to eliminate one or more of the answer choices as wrong, however, guessing is advisable.

What Your Mathematics Test Score Means

Your Mathematics Test score is reported on a scale ranging from 200 to 800. In other words, the lowest mark anyone can possibly get is 200, the highest 800. Your test result will be sent to your high school and to the college (or colleges) which you designate.

The test score is generally reduced to a percentile ranking. The one percent of the test takers that gets the best score on a test is in the 99th percentile; the group that ranks one-fourth of the way from the top in the 75th percentile; the group that ranks in the middle in the 50th percentile; and the group inferior to 90 percent of the applicants in the 10th percentile. For many tests these norms are based on national averages or regional averages, like the New England states or the Mid-Western states. On most college entrance tests, norms are determined and published several months after the college year begins and are based on the experience of all colleges. Since these tests are very similar from year to year, an admissions board can easily determine the relative standing of any candidate immediately after he or she takes the test.

Applying for the Examination

APPLICATION AND REGISTRATION

Every candidate is required to file a formal application with the College Entrance Examination Board, and to pay an examination fee. Write to: College Board SAT Program, Box 6200, Princeton, NJ 08541-6200 for information on application procedures.

ADMISSION TICKETS

After registering, you will be sent a ticket of admission giving the address of the place to which you should report for assignment to an examination room. Do not expect to receive your ticket until approximately one month before the examination date. You will be required to show your ticket to the supervisor at the examination. Normally, no candidate will be admitted to the examination room without a ticket of admission.

A candidate who loses this ticket should immediately write or wire the issuing office for a duplicate authorization.

Rules of Conduct

No books, compasses, rulers, dictionaries, or papers of any kind may be taken into the examination room; you are urged not to bring them to the center at all. Supervisors will not permit anyone found to have such materials to continue a test. Students who will be taking either the Mathematics Level IC or Mathematics Level IIC Examination must bring their own scientific or graphing calculators to the examination. An ordinary four-function calculator will not be sufficient for either the Level IC or IIC exam. Your calculator should be battery or solar powered, not dependent upon an electrical outlet. It may not have printout capability and must be silent. Be certain that your calculator is in good condition; if you bring a backup, become thoroughly familiar with its operation before you come to the exam. No calculators will be provided at the exam site.

Anyone giving or receiving any kind of assistance during the test will be asked to leave the room. The testbook and answer sheet will be taken from the student and returned to CEEB. The answer sheet will not be scored, and the incident will be reported to the institutions designated to receive the score report.

Scratch work may be done in the margins of the testbooks. The use of scratch paper is not permitted. You must turn in all testbooks and answer sheets. Documents or memoranda of any sort are not to be taken from the room.

If you wish to leave the room during a test period or during a test, you must secure permission from the supervisor.

The examinations will be held only on the day and at the time scheduled. Be on time. Under no circumstances will supervisors honor requests for a change in schedule. You will not be permitted to continue a test or any part of it beyond the established time limit. You should bring a watch, but not one with an audible alarm.

To avoid errors or delay in reporting scores:

1. Always use the same form of your name on your application form, answer sheets, and on any correspondence with CEEB. Do not write "John T. Jones, Jr." at one time, and "J. T. Jones" at another. Such inconsistency makes correct identification of papers difficult.

2. Write legibly at all times.

Transmitting the Results

The colleges that you designate receive a report of your scores directly from CEEB. You may have your scores reported to as many as three colleges without additional fee if you designate them in the appropriate place on your application.

After registration closes, you may not substitute or delete institutions already listed on your application. No partial reports will be issued; reports will include scores made on all tests taken on a given date. To avoid duplication of requests, you should keep a record of the institutions to which you have requested that scores be sent.

Score reports requested on the application or by letter before the closing date will be issued within five weeks after your examination date. Although score reports requested after the closing date cannot be sent as quickly, they will be issued as soon as possible.

PART TWO
Outline of Topics

CONTENTS

I. Arithmetic 11
II. Algebra 11
III. Geometry 12
IV. Functions and Their Graphs 13
V. Real Number System 14
VI. Logic 14
VII. Sets 15
VIII. Trigonometry 15
IX. Miscellaneous Topics 16

I. ARITHMETIC

A. Whole numbers
1. Operations—addition, subtraction, multiplication, division
2. Prime and composite numbers
3. Factors and divisors

B. Fractions
1. Types—proper, improper, mixed numbers
2. Operations

C. Decimals
1. Operations
2. Conversions
   a) Decimals to fractions
   b) Fractions to decimals
3. Rounding and approximation

4. Powers of 10
   a) Multiplication
   b) Division
   c) Scientific notation

D. Percent
1. Conversions
   a) Percent to decimal
   b) Decimal to percent
2. Percent problems

E. Ratio and proportion

F. Square roots

G. Averages

H. Metric measurement

II. ALGEBRA

A. Signed numbers
1. Absolute value
2. Inequality and order of signed numbers
3. Addition, subtraction, multiplication, division
4. Order of operations
5. Grouping symbols
6. Evaluating algebraic expressions and formulas

B. Properties of operations
1. Commutative properties
2. Associative properties
3. Distributive properties
4. Special properties of zero
5. Special properties of one
6. Additive and multiplicative inverses

C. Operations with polynomials
1. Exponents and coefficients
2. Addition and subtraction
3. Multiplication
4. Division

D. Equations in one variable
1. Methods of solution
2. Literal equations

E. Inequalities in one variable

F. Systems of equations and inequalities in two variables

G. Verbal Problems
   1. Number
   2. Consecutive integer
   3. Motion
   4. Coin
   5. Mixture
   6. Age
   7. Work
   8. Variation—direct and inverse

H. Special products and factoring
   1. Common monomial factors
   2. Trinomials of the form $ax^2 + bx + c$
   3. Difference of two squares
   4. Complete factoring

I. Algebraic fractions
   1. Reducing fractions
   2. Multiplication
   3. Division
   4. Addition and subtraction
       a) Same denominators
       b) Different denominators
   5. Complex fractions
   6. Equations involving fractions

J. Radicals and irrational numbers
   1. Simplifying radicals
   2. Addition and subtraction of radicals
   3. Multiplication and division of radicals
   4. Rationalizing denominators
   5. Radical equations
   6. Fractional exponents

K. Solution of quadratic equations
   1. Factoring
   2. Completing the square
   3. Formula

L. Graphing
   1. Ordered pairs in the plane
   2. Methods of graphing linear equations
       a) Pairs in the solution set
       b) Intercepts
       c) Slope and slope-intercept method
   3. Parallel and perpendicular lines
   4. Graphing inequalities
   5. Graphical solution of systems of equations

M. Solution of simple cubic equations
   1. Factor theorem
   2. Remainder theorem
   3. Synthetic division
   4. Irrational and complex roots
   5. Solving simple cubic equations

III. GEOMETRY

A. Angles
   1. Types—acute, right, obtuse
   2. Complements and supplements
   3. Vertical angles

B. Lines
   1. Parallel lines and their angles
   2. Perpendicular lines

C. Triangles
   1. Sum of the angles
   2. Congruent triangles
   3. Similar triangles
   4. Special triangles
       a) Isosceles
       b) Equilateral
       c) Right (Pythagorean Theorem)
   5. Vectors

D. Polygons
  1. Quadrilaterals
     a) Parallelogram
     b) Rectangle
     c) Square
     d) Rhombus
     e) Trapezoid
  2. Regular polygons

E. Circles
  1. Special lines and their related angles
     a) Radius and diameter
     b) Chord
     c) Tangent
     d) Secant
  2. Angle and arc measurement
  3. Polygons inscribed in circles

F. Perimeter and area
  1. Triangles
  2. Polygons
  3. Circles
     a) Circumference and arc length
     b) Area of sectors and segments

G. Volume
  1. Pyramid
  2. Prism
  3. Cylinder
  4. Cone
  5. Sphere
  6. Cube
  7. Rectangular solid

H. Coordinate geometry
  1. Coordinate representation of points
  2. Distance between two points
  3. Midpoint of a line segment
  4. Slope of a line
  5. Parallel and perpendicular lines

I. Basic trigonometry
  1. Definition of sine, cosine, tangent
  2. Trigonometry in special triangles
     a) 30°–60°–90° triangle
     b) Isosceles right triangle
  3. Trigonometric problems
     a) Angle of elevation
     b) Angle of depression

IV. FUNCTIONS AND THEIR GRAPHS

A. Relations and functions
  1. Ordered pairs
  2. Function notation
  3. Domain and range
  4. One-to-one functions
  5. Inverse functions
  6. Combining functions
     a) Addition, subtraction, multiplication, division
     b) Composition

B. Graphs
  1. Linear
     a) Slope
     b) Intercepts
  2. Special functions
     a) Absolute value function
     b) Step functions
  3. Polynomial and rational functions
     a) Quadratic—parabola
        i. Axis of symmetry
        ii. Vertex
     b) Cubics
     c) Hyperbola of the form $xy = k$
  4. Related non-function graphs
     a) Circle
     b) Ellipse
     c) Hyperbola of the form $ax^2 - by^2 = c$
  5. Graphs of inverse functions

V. REAL NUMBER SYSTEM

A. Subsets of the real numbers
 1. Natural numbers
    a) Primes
    b) Composites—prime factorization
 2. Integers
    a) Multiples and divisors
       i. Factors
       ii. Divisibility
       iii. Least common multiple
       iv. Greatest common divisor
       v. Perfect squares
    b) Odd and even integers
 3. Rational and irrational numbers
    a) Decimal representations
    b) Simplification of radicals and exponents
    c) Identifying rational and irrational numbers

B. Operations and properties
 1. Properties of the binary operations
    a) Closure
    b) Commutative properties
    c) Associative properties
    d) Distributive properties
 2. Absolute value
 3. Real number line
    a) Order
    b) Density
    c) Completeness
 4. Properties of zero and one
    a) Identity elements
    b) Additive and multiplicative inverses
    c) Division involving zero
    d) Zero as an exponent
 5. Nature of the roots of quadratic equations
 6. Pythagorean triples

VI. LOGIC

A. Propositions
 1. Simple statements
    a) Symbols
    b) Quantifiers (all, some)
 2. Negation
 3. Compound statements
    a) Conjunction
    b) Disjunction
    c) Implication (conditional statements)
       i. Necessary conditions
       ii. Sufficient conditions
       iii. Equivalence (necessary and sufficient conditions)
    d) Derived implications
       i. Converse
       ii. Inverse
       iii. Contrapositive

B. Truth tables

C. Methods of proof
 1. Valid arguments
    a) Direct
    b) Indirect—contradiction and counterexample
 2. Invalid arguments—fallacies

VII. SETS

A. Meaning and symbols
   1. Set notation
   2. Set membership
   3. Ordered pairs
   4. Cardinality of a set

B. Types of sets
   1. Finite
   2. Infinite
   3. Empty

C. Relationships between sets
   1. Equal sets
   2. Equivalent sets
   3. Subsets
   4. Complements

D. Set Operations
   1. Union
   2. Intersection
   3. Cartesian products
   4. Laws of set operations
   5. Closure

E. Venn diagrams

VIII. TRIGONOMETRY

A. Trigonometry of the right triangle
   1. Definitions of the six functions
   2. Relations of the functions of the complementary angles
   3. Reciprocal relations among the functions
   4. Variations in the functions of acute angles
   5. Pythagorean and quotient relations
   6. Functions of 30°, 45° and 60°
   7. Applications of the functions to right triangle problems

B. Trigonometric functions of the general angle
   1. Generating an angle of any size
   2. Radians and degrees
   3. Using radians to determine arc length
   4. Definitions of the functions of an angle
   5. Signs of the functions in the four quadrants
   6. Functions of the quadrantal angle
   7. Finding the value of functions of any angle

C. Identities and equations
   1. Difference between identities in equations
   2. Proving identities
   3. Solving linear trigonometric functions
   4. Solving trigonometric quadratic equations

D. Generalized trigonometric relationships
   1. Functions of the sum of two angles
   2. Functions of the difference of two angles
   3. Functions of the double angle
   4. Functions of the half angle

E. Graphs of trigonometric functions
   1. Graphs of the sine, cosine and tangent curves
   2. Properties of the sine, cosine and tangent curves
   3. Definitions of amplitude, period and frequency
   4. Solving trigonometric equations graphically

F. Solutions of oblique triangles
   1. Law of sines
   2. Law of cosines
   3. Using logarithms to solve oblique triangle problems
   4. Vector problems—parallelogram of forces
   5. Navigation problems

IX. MISCELLANEOUS TOPICS

A. Complex numbers
1. Meaning
2. Operations
   a) Addition and subtraction
   b) Multiplication and division
      i. Powers of i
      ii. Complex conjugate
3. Complex roots of quadratic equations

B. Number Bases
1. Converting from base 10 to other bases
2. Converting from other bases to base 10
3. Operations in other bases

C. Exponents and logarithms
1. Meaning of logarithms
2. Computation with exponents and logarithms
3. Equations
4. Graphs of exponential and logarithmic functions

D. Binary operations
1. Definition of binary operations
2. Properties of binary operations
3. Application to modular arithmetic

E. Identity and inverse elements
1. Addition
2. Multiplication
3. Other operations

PART THREE

Math Review—Illustrative Problems and Solutions

CONTENTS

1. Formulas and Linear Equations 19
2. Algebraic Fractions 22
3. Sets 24
4. Functions 25
5. Exponents 28
6. Logarithms 31
7. Equations—Quadratic, Radical and Exponential 33
8. Inequalities 37
9. Verbal Problems 39
10. Geometry 42
11. Trigonometry 56
12. Graphs and Coordinate Geometry 64
13. Number Systems and Concepts 67
14. Arithmetic and Geometric Progressions 71
15. Vectors 74
16. Variation 76

1. Formulas and Linear Equations

An *equation* is a statement that two mathematical expressions are equal.

In the equation $3x + 4 = 19$, the 3, 4, and 19 are called *constants*, the letter x the *variable*. When solving an equation we try to find the numerical value (or values) of the variable that makes the equality true. In $3x + 4 = 19$, the value $x = 5$ is the *root* or *solution* of the equation. In this equation the highest exponent of x is 1, and so we call such an equation a *first degree* equation. It is also called a *linear* equation, since its graph is a *straight line*.

The basic principle of solving equations is the following:

Addition, subtraction, multiplication, or division (except by 0) of *both* sides of an equation by the same number results in an equivalent equation, i.e., one with the same root or roots.

To solve $3x + 4 = 19$, start by subtracting 4 from both sides.

$$\begin{aligned} 3x + 4 &= 19 \\ -4 &= -4 \\ \hline 3x &= 15 \end{aligned}$$

Now divide both sides by 3.

$$\frac{3x}{3} = \frac{15}{3}$$
$$x = 5$$

To solve *fractional equations*, first multiply both sides of the equation by the least common denominator (LCD) of all fractions in the equation.

To solve $\frac{y}{3} - \frac{y}{5} = 4$, multiply both sides of the equation by 15, the least common denominator (LCD).

$$\frac{y}{3}(15) - \frac{y}{5}(15) = 4(15)$$
$$5y - 3y = 60$$
$$2y = 60$$
$$\frac{2y}{2} = \frac{60}{2}$$
$$y = 30$$

Substitution of 30 for y in the original equation serves as a check of the answer.

A *formula* is an equation usually involving a relationship between literal quantities. Problems involving formulas often require substitution in a formula and solution of the resulting equation for a particular variable.

If the formula is entirely literal and the problem calls for solving for one variable in terms of the others, start by moving all terms containing this variable to one side of the equation.

The area, A, of a triangle is given by the formula:

$$A = \frac{1}{2}bh \quad (b = \text{base}, \ h = \text{altitude})$$

To solve for h, multiply both sides by 2.

$$\frac{2A}{b} = \frac{bh}{b}$$
$$h = \frac{2A}{b}$$

ILLUSTRATIVE PROBLEMS

1. In the formula $F = \frac{9}{5}C + 32$, find C when $F = 68$.

Solution▶ Substitute 68 in the formula.

$$68 = \frac{9}{5}C + 32$$

Subtract 32 from both sides.

$$36 = \frac{9}{5}C$$

Multiply both sides by 5.

$$180 = 9C$$
$$C = 20$$

2. Solve the formula $s = \frac{at}{a+t}$ for t.

Solution▶ Multiply both sides by $a + t$.

$$s(a+t) = at$$
$$as + st = at$$

Subtract st from both sides.

$$as = at - st$$

Factor the right side.

$$as = (a-s)t$$
$$t = \frac{as}{a-s}$$

3. In the formula $V = \pi r^2 h$, if r is doubled, what must be done to h to keep V constant?

Solution▶ If r is doubled, the effect is to quadruple V, since the r is squared in the formula. Hence, h must be divided by 4 to keep V the same in value.

Solution▶ 4. A package weighing 15 lb is sent by parcel post. It costs x cents for the first 10 lb and y cents for each additional lb. Express the cost, C, in terms of x and y.

The first 10 lb cost x cents; the remaining 5 lb cost $5y$ cents. The total cost C is given by the formula:

$$C = x + 5y$$

5. Solve for m: $2m + 7 = m - 9$

Solution▶ Subtract m and 7 from both sides.

$$\begin{aligned} 2m + 7 &= m - 9 \\ -m - 7 &= -m - 7 \\ \hline m &= -16 \end{aligned}$$

6. Solve for y: $\dfrac{y}{4} + \dfrac{y}{3} = \dfrac{7}{12}$

Solution▶ Multiply both sides by 12 (LCD).

$$\frac{y}{4}(12) + \frac{y}{3}(12) = \frac{7}{12}(12)$$
$$3y + 4y = 7$$
$$7y = 7$$
$$y = 1$$

7. Solve for n: $an = 5 + bn$

Solution▶ Subtract bn from both sides.

$$\begin{aligned} an &= 5 + bn \\ -bn &= - bn \\ \hline an - bn &= 5 \end{aligned}$$

Now factor on the left side.

$$n(a - b) = 5$$

Divide both sides by $(a - b)$.

$$\frac{n(a-b)}{(a-b)} = \frac{5}{a-b}$$

$$n = \frac{5}{a-b}$$

2. Algebraic Fractions

To reduce or multiply algebraic fractions with binomial or polynomial terms, first factor the polynomial completely, and then cancel out factors that are common to both numerator and denominator of the fraction or fractions.

To divide algebraic fractions, invert the divisor and proceed as in multiplication.

To add or subtract algebraic fractions, convert the fractions to equivalent fractions with the same least common denominator (LCD) and then add like fractions as in adding arithmetic fractions.

In the following illustrative problems, we assume that the variables do not take values which make the denominator zero.

ILLUSTRATIVE PROBLEMS

1. Reduce to lowest terms: $\dfrac{y^2 - 5y}{y^2 - 4y - 5}$

Solution   Factor numerator and denominator.

$$\frac{y(y-5)}{(y+1)(y-5)}$$

Divide numerator and denominator by $(y-5)$.

$$\frac{\cancel{y-5}}{\cancel{y-5}} \cdot \frac{y\cancel{(y-5)}}{(y+1)\cancel{(y-5)}} = \frac{y}{y+1}$$

2. Multiply: $\dfrac{x^2}{xy - y^2} \cdot \dfrac{x^2 - y^2}{x^2 + xy}$

Solution   Factor numerators and denominators.

$$\frac{x^2}{y(x-y)} \cdot \frac{(x+y)(x-y)}{x(x+y)}$$

Divide numerators and denominators by common factors x, $(x-y)$, $(x+y)$.

$$\frac{x \cdot \cancel{x}}{y\cancel{(x-y)}} \cdot \frac{\cancel{(x+y)}\cancel{(x-y)}}{\cancel{x}\cancel{(x+y)}} = \frac{x}{y}$$

3. Divide: $\dfrac{b+5}{b} \div \dfrac{b^2-25}{b^2}$

Solution▸ Invert second fraction and multiply.

$$\frac{b+5}{b} \cdot \frac{b^2}{b^2-25} = \frac{\cancel{b+5}}{b} \cdot \frac{\cancel{b}\cdot b}{\cancel{(b+5)}(b-5)}$$

$$= \frac{b}{(b-5)}$$

4. If a man buys several articles for n cents per dozen and his selling price is $\dfrac{n}{9}$ cents per article, what is his profit, in cents, on each article?

Solution▸

$$\text{Profit} = \text{S.P.} - \text{Cost}$$

$$\text{Profit} = \frac{n}{9} - \frac{n}{12}$$

Lowest common denominator is 36.

$$\text{Profit} = \frac{4n}{36} - \frac{3n}{36}$$

$$= \frac{n}{36}$$

5. Simplify: $\left(\dfrac{1}{a} - \dfrac{1}{b}\right) \div \left(1 - \dfrac{a}{b}\right)$

Solution▸ Convert each expression in parentheses into a single fraction.

$$\frac{b-a}{ab} \div \frac{b-a}{b} = \frac{\cancel{b-a}}{ab} \cdot \frac{\cancel{b}}{\cancel{b-a}}$$

$$= \frac{1}{a}$$

6. Simplify: $\dfrac{3 + \dfrac{1}{x}}{9 - \dfrac{1}{x^2}}$

Solution▸ Multiply numerator and denominator by x^2.

$$\frac{3x^2 + x}{9x^2 - 1} = \frac{x\cancel{(3x+1)}}{(3x-1)\cancel{(3x+1)}}$$

$$= \frac{x}{3x-1}$$

7. Simplify: $\dfrac{\sin^4 A - \cos^4 A}{\sin^2 A - \cos^2 A}$

Solution Factor the numerator.

$$\dfrac{(\sin^2 A + \cos^2 A)(\cancel{\sin^2 A - \cos^2 A})}{(\cancel{\sin^2 A - \cos^2 A})}$$

$$\sin^2 A + \cos^2 A = 1$$

8. If $y = 1 - \dfrac{1}{x}$, $x > 0$, what effect does an increase in x have on y?

Solution As x increases, $\dfrac{1}{x}$ decreases. Therefore, we are subtracting a smaller quantity from 1, and consequently y increases.

3. Sets

The *solution set* of an open sentence is the set of all elements in the replacement set of the variable that make the open sentence a true sentence.

The *intersection* of two sets P and Q ($P \cap Q$) is the set of all elements that are members of *both* P and Q.

The *union* of two sets P and Q ($P \cup Q$) is the set of all elements that are members of *either* P or Q.

Two sets are said to be *disjoint* sets when their intersection is the *empty* set. ($P \cap Q = \emptyset$)

The *complement* of a set P is the set P' of all members of the universal set which are not members of P.

ILLUSTRATIVE PROBLEMS

1. How many elements are in the set: $\{x \mid 3 < x < 9, x \text{ is an integer}\}$?

Solution The indicated set contains only the elements 4, 5, 6, 7, and 8, or 5 elements.

2. If A is the set of all prime numbers and B the set of all even integers, what set is represented by $A \cap B$?

Solution The only even prime integer is 2.

$$A \cap B = \{2\}$$

3. Find the solution set of the equation $x^2 = 3x$ if x is the set of real numbers.

Solution▶
$$x^2 - 3x = 0$$
$$x(x-3) = 0$$
$$x = 0 \text{ or } x - 3 = 0$$
$$x = 3$$

The solution set is $\{0, 3\}$

4. Find the solution set of $3x - 4 > x + 2$ where x is the set of the real numbers.

Solution▶
$$3x - 4 > x + 2$$
$$3x - x > 4 + 2$$
$$2x > 6$$
$$x > 3$$

5. Find the solution set of the system: $A = \{(x, y) \mid x^2 + y^2 = 25\}$ and $B = \{(x, y) \mid y = x + 1\}$

Solution▶ Substitute $y = x + 1$ into the first equation.

$$x^2 + (x+1)^2 = 25$$
$$x^2 + x^2 + 2x + 1 = 25$$
$$2x^2 + 2x - 24 = 0$$
$$x^2 + x - 12 = 0$$
$$(x+4)(x-3) = 0$$

Thus $x + 4 = 0$ or $x - 3 = 0$ so that $x = -4$ or $x = 3$.
When $x = -4$, $y = -3$ and when $x = 3$, $y = 4$.
$A \cap B$ has two elements: $(3, 4)$ and $(-4, -3)$

4. Functions

A *function* is a set of ordered pairs (x, y) such that for each value of x, there is one and only one value of y. We then say that "y is a function of x," written $y = f(x)$ or $y = g(x)$, etc. The set of x-values for which the set is defined is called the *domain* of the function, and the set of corresponding values of y is called the *range* of the function.

y is said to be a *linear* function of x if the two variables are related by a first-degree equation, such as $y = ax + b$ where $a \neq 0$ and b is any real number.

y is said to be a *quadratic* function of x if y can be expressed in the form $y = ax^2 + bx + c$ where $a \neq 0$ and b and c are real numbers.

In general, y is said to be a *polynomial* function of x if y can be expressed in the form:

$$y = c_0 x^n + c_1 x^{n-1} + c_2 x^{n-2} + \ldots + c_{n-1} x + c_n$$

where the exponents are non-negative integers and the coefficients ($c_0, c_1, c_2, \ldots c_n$) are real numbers.

When we speak of $f(a)$, we mean the value of $f(x)$ when $x = a$ is substituted in the expression for $f(x)$.

The *inverse* of a function is obtained by interchanging x and y in the equation $y = f(x)$ that defines the function. The inverse of a function may or may not be a function. A procedure that is often used to find the inverse of a function $y = f(x)$ is to interchange x and y in the equation that relates them, and then to solve for y in terms of x, if possible.

If $z = f(y)$ and $y = g(x)$, we may say that $z = f[g(x)]$. Thus z is in turn a function of x. In this case we may say that z is a composite function of f and g and is also written $f \cdot g = f[g(x)]$. For example, if $z = f(y) = 3y + 2$ and $y = g(x) = x^2$, then $z = f[g(x)] = 3[g(x)] + 2 = 3x^2 + 2$.

ILLUSTRATIVE PROBLEMS

1. If $f(x) = x^2 + 2x - 5$, find the value of $f(2)$.

Solution Substitute $x = 2$ in the polynomial.

$$2^2 + 2(2) - 5 = 4 + 4 - 5 = 3$$

2. If $f(y) = \tan y + \cot y$, find the value of $f\left(\frac{\pi}{4}\right)$.

Solution
$$f\left(\frac{\pi}{4}\right) = \tan \frac{\pi}{4} + \cot \frac{\pi}{4}$$
$$= 1 + 1$$
$$= 2$$

3. If $F(t) = t^2 + 1$, find $F(a - 1)$.

Solution Substitute $t = a - 1$.

$$F(a - 1) = (a - 1)^2 + 1$$
$$= (a^2 - 2a + 1) + 1$$
$$= a^2 - 2a + 2$$

4. If $f(x) = 2x + 3$ and $g(x) = x - 3$, find $f[g(x)]$.

Solution In $f(x)$, substitute $g(x)$ for x.

$$f[g(x)] = 2[g(x)] + 3$$
$$= 2(x - 3) + 3$$
$$= 2x - 6 + 3$$
$$= 2x - 3$$

5. What are the *domain* and *range* of the function $y = |x|$?

Solution▶ The function is defined for all real values of x. Hence the *domain* is $\{x \mid -\infty < x < +\infty; x \text{ is a real number}\}$.

Since $y = |x|$ can only be a positive number or zero, the *range* of the function is given by the set $\{y \mid 0 \leq y < +\infty; y \text{ is a real number}\}$.

6. If $f(t) = \dfrac{1+t}{t}, f(t) =$

(A) $+f(-t)$ (B) $f\left(\dfrac{1}{t}\right)$ (C) $f\left(-\dfrac{1}{t}\right)$ (D) $\dfrac{1}{t}f\left(\dfrac{1}{t}\right)$ (E) none of these

(A) $f(-t) = \dfrac{1-t}{-t} = \dfrac{t-1}{t} \neq f(t)$

(B) $f\left(\dfrac{1}{t}\right) = \dfrac{1+\dfrac{1}{t}}{\dfrac{1}{t}} = t+1 \neq f(t)$

(C) $f\left(-\dfrac{1}{t}\right) = \dfrac{1-\dfrac{1}{t}}{-\dfrac{1}{t}} = \dfrac{-t+1}{1} \neq f(t)$

(D) $\dfrac{1}{t}f\left(\dfrac{1}{t}\right) = (t+1) \cdot \dfrac{1}{t}$ (from (B))

$\dfrac{1}{t}f\left(\dfrac{1}{t}\right) = \dfrac{t+1}{t} + f(t)$

7. Find the largest real range of the function $y = 1 - \dfrac{1}{x}$.

Solution▶ Solve for $\dfrac{1}{x} = 1 - yx$:

$$x = \dfrac{1}{1-y}$$

The range for y consists of all real numbers except $y = 1$.

8. Write the inverse of the function f as defined by $f(x) = \sqrt{x-1}$.

Solution▶ Let $y = \sqrt{x-1}$. Substitute x for y, and y for x.

$$x = \sqrt{y-1}$$
$$y = x^2 + 1$$
$$f^{-1}(x) = x^2 + 1$$

28 SAT II: Math

9. If $f(x)=\dfrac{x}{x-1}$, $f(x+1)=$

Solution▶

$$f(x+1)=\dfrac{x+1}{x+1-1}$$

$$=\dfrac{x+1}{x}=1+\dfrac{1}{x}$$

$$\dfrac{1}{f(x)}=\dfrac{x-1}{x}=1-\dfrac{1}{x}$$

Hence $f(x+1)=\left(1-\dfrac{1}{x}\right)+\dfrac{2}{x}$

$$=\dfrac{1}{f(x)}=\dfrac{2}{x}$$

10. If the functions f and g are defined as $f(x)=x^2-2$ and $g(x)=2x+1$, what is $f[g(x)]$?

Solution▶

$$f[g(x)]=[g(x)]^2-2$$
$$=(2x+1)^2-2$$
$$=4x^2+4x+1-2$$
$$=4x^2+4x-1$$

5. Exponents

The following formulas and relationships are important in solving problems dealing with exponents ($x \neq 0$ in all cases that follow):

$$x^0=1$$

$$x^{-n}=\dfrac{1}{x^n}$$

$$x^{m/n}=\sqrt[n]{x^m}=\left(\sqrt[n]{x}\right)^m \text{ where } m \text{ and } n \text{ are integers, } n \neq 0$$

$$x^a \cdot x^b = x^{a+b}$$

$$\dfrac{x^a}{x^b}=x^{a-b}$$

$$(x^a)^b = x^{ab}$$

$$(xy)^a = x^a y^a$$

In *scientific notation* a number is expressed as the product of a number between 1 and 10 and an integral power of 10. This notation provides a convenient way of writing very large or very small numbers and simplifies computation with such numbers. For example, if a certain star is 780 billion miles from the earth, we write this number as 7.8×10^{11}. The eleventh power of 10 indicates that the decimal point in 7.8 is to be moved 11 places to the right.

If the diameter of a certain atom is .00000000092 cm., we write this number as 9.2×10^{-10}. The -10, as a power of 10, indicates that the decimal point is to be moved 10 places to the left.

This method of writing large and small numbers is consistent with the laws of exponents above. These laws also facilitate computation with very large or very small numbers when written in scientific notation, as illustrated in some of the problems below.

ILLUSTRATIVE PROBLEMS

1. Find the value of $2x^0 + x^{2/3} + x^{-2/3}$ when $x = 27$.

Solution Substitute $x = 27$.

$$2(27)^0 + (27)^{2/3} + (27)^{-2/3}$$
$$= 2 \cdot 1 + \left(\sqrt[3]{27}\right)^2 + \frac{1}{27^{2/3}}$$
$$= 2 + 9 + \frac{1}{9}$$
$$= 11\frac{1}{9}$$

2. If $y = 3^x$, $3^{x+2} =$

(A) y^2 (B) 2^y (C) $y + 3$ (D) $9y$ (E) $y + 9$

Solution
$$3^{x+2} = 3^x \cdot 3^2$$
$$= y \cdot 9$$
$$= 9y$$

3. If 0.00000784 is written in the form 7.84×10^n, what does n equal?

Solution Writing the number in scientific notation, we get $0.00000784 = 7.84 \times 10^{-6}$.

$$n = -6$$

4. The length of an electromagnetic wave is given by the formula $L = \frac{C}{F}$, where C is the velocity of light ($3^5 \times 10^{10}$ cm per sec) and F is the frequency. What is the value of L when $F = 3000$ megacycles per sec?

Solution $F = 3000 \times 10^6 = 3 \times 10^9$

Substitute in formula.

$$L = \frac{3 \times 10^{10}}{3 \times 10^9} = 1 \times 10^1$$
$$= 10 \text{ cm}$$

30 SAT II: Math

5. Solve the exponential equation: $3^{2x-1} = 81$

Solution>
$$3^{2x-1} = 3^4$$

Since the bases are equal, equate the exponents.

$$2x - 1 = 4$$
$$2x = 5$$
$$x = 2\frac{1}{2}$$

6. If $4^y = 125$, between what two consecutive integers does y lie?

Solution>
$$4^3 = 64$$
$$4^4 = 256$$

Since 125 is between 64 and 256 and 4^y is a steadily increasing function, y is between 3 and 4.

7. Solve the equation: $9^{x+3} = \frac{1}{27}$

Solution>
$$\left(3^2\right)^{x+3} = \frac{1}{3^3} = 3^{-3}$$
$$3^{2x+6} = 3^{-3}$$

Since the bases are equal, the exponents may be set equal.
$$2x + 6 = -3$$
$$2x = -9$$
$$x = -4\frac{1}{2}$$

8. Solve for x: $2^{x+2} = \left(\frac{1}{2}\right)^x$

Solution>
$$2x + 2 = 2^{-x}$$
$$x + 2 = -x$$
$$2x = -2$$
$$x = -1$$

9. Solve for r: $27^{6-r} = 9^{r-1}$

(A) 1   (B) 2   (C) 3   (D) 4   (E) 5

Solution>

(D) $3^{3(6-r)} = 3^{2(r-1)}$

If the bases are equal, the exponents are equal.
$$3(6 - r) = 2(r - 1)$$
$$18 - 3r = 2r - 2$$
$$20 = 5r$$
$$r = 4$$

10. Find the value, in simplest form, of the fraction $\dfrac{2^{n+4} - 2(2^n)}{2(2^{n+3})}$.

(A) $\dfrac{1}{2}$ (B) $\dfrac{1}{4}$ (C) $\dfrac{(3)}{(4)}$ (D) $\dfrac{5}{8}$ (E) $\dfrac{7}{8}$

Solution▶ (E)

$$\frac{2^{n+4} - 2^{n+1}}{2^{n+4}} = 1 - 2^{-3}$$

$$= 1 - \frac{1}{2^3} = 1 - \frac{1}{8}$$

$$= \frac{7}{8}$$

6. Logarithms

Definition: The *logarithm* of a number to a given base is the exponent to which this base must be raised to yield the given number.

$\log_b N = e$ is equivalent to $b^e = N$. For example, the equation $5^3 = 125$ may be written $\log_5 125 = 3$. For computational purposes we usually use 10 as the base; if we write $\log n$, the base 10 is understood. Logarithms to base 10 are called *common* logarithms.

The function inverse to the function $y = b^x$, $b > 0$, $b \neq 1$ is $y = \log_b x$. We define the logarithmic function of x at this point only for positive values of x.

The laws of logarithms are derived from the laws of exponents. They are listed below for base 10 although they apply to any acceptable base.

$$\log(ab) = \log a + \log b$$

$$\log \frac{a}{b} = \log a - \log b$$

$$\log a^n = n \cdot \log a$$

$$\log a^{1/n} = \log \sqrt[n]{a} = \frac{1}{n} \log a$$

Although common logarithms are generally used for computation, logarithms to base e are used in more advanced work, particularly in calculus. The constant $e = 2.7183 \ldots$ is an irrational number, and is significant in the study of organic growth and decay. The function $y = e^x$ is usually called the *exponential* function.

ILLUSTRATIVE PROBLEMS

1. Find the value of $\log_4 64$.

Solution▶ Let $x = \log_4 64$. In exponential notation,

$$4^x = 64$$
$$x = 3$$

2. If log 63.8 = 1.8048, what is log 6.38?

Solution

$$\log 6.38 = \log \frac{63.8}{10}$$
$$= \log 63.8 - \log 10$$
$$= 1.8048 - 1$$
$$= 0.8048$$

3. If log 2 = a and log 3 = b, express log 12 in terms of a and b.

Solution

$$\log 12 = \log(4 \cdot 3)$$
$$= \log 4 + \log 3$$
$$= \log 2^2 + \log 3$$
$$= 2 \log 2 + \log 3$$
$$= 2a + b$$

4. In the formula $A = P(1 + r)^n$, express n in terms of A, P, and r.

Solution

$$\frac{A}{P} = (1+r)^n$$
$$\log \frac{A}{P} = \log(1+r)^n = n \log(1+r)$$
$$\log A - \log P = n \log(1+r)$$
$$n = \frac{\log A - \log P}{\log(1+r)}$$

5. If $\log t^2 = 0.8762$, log 100t =

Solution

$$2 \log t = 0.8762$$
$$\log t = 0.4381$$
$$\log 100t = \log 100 + \log t$$
$$= 2 + 0.4381$$
$$= 2.4381$$

6. If log tan x = 0, find the smallest positive value of x.

If log tan x = 0, then tan x = 1.

Therefore $x = \frac{\pi}{4}$

7. If $\log_a 2 = x$ and $\log_a 5 = y$, express $\log_a 40$ in terms of x and y.

Solution

$$\log 40 = \log(8 \cdot 5) = \log 8 + \log 5$$
$$= \log 2^3 + \log 5$$
$$= 3 \log 2 + \log 5$$
$$= 3x + y$$

8. Find $\log_3 3\sqrt{3}$.

(A) 3   (B) 1   (C) $\frac{3}{2}$ (D) $\frac{2}{3}$ (E) none of these

Solution ▶ (C)
$$\log_3 3\sqrt{3} = \log_3 3^{3/2}$$
$$= \frac{3}{2} \log_3 3$$
$$= \frac{3}{2}(1) = \frac{3}{2}$$

7. Equations—Quadratic, Radical and Exponential

An equation of the second degree is called a *quadratic* equation. The general form of the quadratic equation in one variable is

$ax^2 + bx + c = 0$ where a, b, and c are real numbers and $a \neq 0$.

In order to solve a quadratic equation, express it in its general form, and attempt first to factor the quadratic polynomial. Then set each linear factor equal to zero. This procedure produces *two* roots, which in some cases are equal.

If the quadratic member of the equation $ax^2 + bx + c = 0$ is *not* factorable, we apply the *quadratic formula*:

$$x = \frac{-b \pm \sqrt{b^2 - 4ac}}{2a}$$

The quantity under the radical sign, $b^2 - 4ac$, is called the *discriminant* (D) of the quadratic equation; it determines the nature of the roots of the equation.

If $D = b^2 - 4ac$ is a *positive* quantity, the two roots are real. If D is a *negative* quantity, the roots are *imaginary*. If $D = 0$, the roots are real and equal. If D is a *perfect square*, the roots are real and *rational*.

The roots, r_1 and r_2, of the general quadratic equation $ax^2 + bx + c = 0$ are related to the coefficients of the equation as follows:

$$r_1 + r_2 = -\frac{b}{a} \quad \text{and} \quad r_1 r_2 = \frac{c}{a}$$

An equation containing the variable under a radical sign is called a *radical* equation. In a radical equation both members may be squared, or raised to any power, to eliminate radicals. This procedure may introduce *extraneous* roots; all roots obtained by this method must be checked in the original equation.

When the variable in an equation appears as an exponent in one or more terms, we call the equation an *exponential* equation. One approach in solving such equations is to try to write the terms to the *same base*.

$$2^{y-1} = 8^{-y}$$
$$2^{y-1} = (2^3)^{-y}$$
$$2^{y-1} = 2^{-3y}$$

Since the bases are the same, we may now *equate* the *exponents*.

$$y - 1 = -3y$$
$$4y = 1$$
$$y = \frac{1}{4}$$

In more involved exponential equations it is often helpful to take the *logarithm* of both members.

In solving a pair of *simultaneous equations* in two variables, try to eliminate one of the unknowns and solve for the other.

ILLUSTRATIVE PROBLEMS

1. Find the roots of the equation $x^2 - 5x + 6 = 0$.

Solution Factor the left member.

$$(x - 3)(x - 2) = 0$$
Either $x - 3 = 0$ or $x - 2 = 0$
$$x = 3 \quad \text{and} \quad x = 2$$

2. Solve the following system of equations:

$$x^2 - 3y^2 = 13$$
$$x = 1 - 3y$$

Solution Substitute the value of x from the second equation into the first.

$$(1 - 3y)^2 - 3y^2 = 13$$
$$1 - 6y + 9y^2 - 3y^2 = 13$$
$$6y^2 - 6y - 12 = 0$$
$$y^2 - y - 2 = 0$$
$$(y - 2)(y + 1) = 0$$

$y = 2$	$y = -1$
$x = 1 - 3y$	$x = 1 - 3y$
$x = 1 - 3(2)$	$x = 1 - 3(-1)$
$x = -5$	$x = 4$

Group the roots.

x	-5	4
y	2	-1

7. EQUATIONS—QUADRATIC, RADICAL AND EXPONENTIAL

3. Find the nature of the roots of the equation: $2x^2 - 4x - 3 = 0$.

Solution▸
$$\text{discriminant } (D) = b^2 - 4ac$$
$$D = (-4)^2 - 4(2)(-3)$$
$$D = 16 + 24 = 40$$

Since D is positive but not a perfect square, the roots are real, unequal, and irrational.

4. Solve the equation: $\sqrt{y-2} = 14 - y$

Solution▸ Square both sides.

$$y - 2 = (14 - y)^2$$
$$y - 2 = 196 - 28y + y^2$$
$$y^2 - 29y + 198 = 0$$
$$(y - 11)(y - 18) = 0$$
$$y = 11 \quad \text{and} \quad y = 18$$

It is essential that these roots now be checked in the original equation. Substitute $y = 11$ in the equation.

$$\sqrt{11-2} = 14 - 11 \quad \text{or} \quad \sqrt{9} = 3, \text{ which checks.}$$

Now substitute $y = 18$.

$$\sqrt{18-2} = 14 - 18 \quad \text{or} \quad \sqrt{16} = -4.$$

This value does not check and $y = 11$ is the only root.

5. For what value of a in the equation $ax^2 - 6x + 9 = 0$ are the roots of the equation equal?

Solution▸ Set the discriminant equal to zero.

$$(-6)^2 - 4a \cdot 9 = 0$$
$$36 = 36a$$
$$a = 1$$

6. If 2 is one root of the equation $x^3 - 4x^2 + 14x - 20 = 0$, find the other two roots.

Solution▸ Use synthetic division.

1	−4	14	−20	⌊2
	2	−4	20	
1	−2	10	0	

The resulting equation is $x^2 - 2x + 10 = 0$.

Solve by the quadratic formula.

$$x = \frac{2 \pm \sqrt{4-40}}{2} = \frac{2 \pm 6i}{2}$$

$$x = 1 \pm 3i$$

7. Find K so that 5 is a root of the equation $y^4 - 4y^3 + Ky - 10 = 0$.

Solution▸ Substitute $y = 5$ into the equation.

$$625 - 4(125) + 5K - 10 = 0$$
$$625 - 500 + 5K - 10 = 0$$
$$5K = -115$$
$$K = -23$$

8. Find all positive values of t less than 180° that satisfy the equation $2\sin^2 t - \cos t - 2 = 0$.

Solution▸ Substitute $1 - \cos^2 t$ for $\sin^2 t$.

$$2(1 - \cos^2 t) - \cos t - 2 = 0$$
$$2 - 2\cos^2 t - \cos t - 2 = 0$$
$$2\cos^2 t + \cos t = 0$$
$$\cos t(2\cos t + 1) = 0$$
$$\cos t = 0 \text{ or } 2\cos t + 1 = 0$$

$$t = 90° \quad \Big| \quad \cos t = -\tfrac{1}{2}$$
$$t = 120°$$

$$t = 90°, 120°$$

9. Find the remainder when $x^{16} + 5$ is divided by $x + 1$.

Solution▸ If $f(x) = x^{16} + 5$, then the remainder is equal to

$$f(-1) = (-1)^{16} + 5$$
$$= 1 + 5 = 6$$

10. If two roots of the equation $x^3 + ax^2 + bx + c = 0$ (with a, b, and c integers) are 1 and $2 - 3i$, find the value of a.

Solution▸ Another root must be $2 + 3i$.

$$\text{Sum of the roots} = -a$$
$$1 + (2 + 3i) + (2 - 3i) = -a$$
$$a = -5$$

11. Solve the equation $y^2 = \dfrac{x^2}{x-2}$ for x in terms of y.

Solution▸

$$x = xy^2 - 2y^2$$
$$2y^2 = xy^2 - x$$
$$x(y^2 - 1) = 2y^2$$
$$x = \frac{2y^2}{y^2 - 1}$$

12. How many roots does the equation $\sqrt{x+6} = -x$ have?

Solution

$$x + 6 = x^2$$
$$x^2 - x - 6 = 0$$
$$(x-3)(x+2) = 0$$
$$x = 3, x = -2$$

Check $x = 3, \sqrt{9} = -3$; does not check
Check $x = -2, \sqrt{4} = 2$; does check
There is *one* root.

8. Inequalities

The following *principles* are important in solving problems dealing with *inequalities*.

1. For all real values of p, q, and r, if $p > q$, then $p + r > q + r$.

2. For all real values of p, q, and r ($r \neq 0$), if $p > q$, then $pr > qr$ for values of $r > 0$; and $pr < qr$ for values of $r < 0$.

3. If the $|x| < a$, then $-a < x < a$.

4. The sum of two sides of a triangle is greater than the third side.

5. If two sides of a triangle are unequal, the angles opposite are unequal and the greater angle lies opposite the greater side, and conversely.

In solving quadratic inequalities or trigonometric inequalities, a graphic approach is often desirable.

ILLUSTRATIVE PROBLEMS

1. Find the solution set of the inequality $8y - 5 > 4y + 3$.

Solution Subtract $4y$ from both sides and add 5.

$$8y - 4y > 5 + 3$$
$$4y > 8$$
$$y > 2$$

2. Find the solution set of the inequality $|x + 3| < 5$.

Solution

When $x + 3 > 0$, $|x + 3| = x + 3$
$$x + 3 < 5$$
$$= x < 2$$
When $x + 3 < 0$, $|x + 3| = -(x + 3)$
$$-(x + 3) < 5$$
$$= x + 3 > -5$$
$$x > -8$$

The solution set consists of 1 interval: $-8 < x < 2$.

3. In $\triangle PQR$, $PQ = PR = 5$ and $60° < P < 90°$. What is the possible range of values of QR?

Solution

When $\angle P = 60°$, $\triangle PQR$ is equilateral and $QR = 5$. When $\angle P = 90°$, $\triangle PQR$ is right, isosceles, and $QR = 5\sqrt{2}$

$5 \leq QR \leq 5\sqrt{2}$

4. For what values of x between 0 and $360°$ is $\sin x > \cos x$?

Solution   Graph the two functions on the same set of axes.

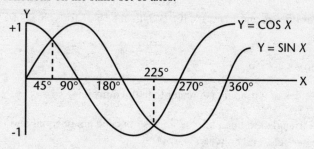

From the graph it is apparent that $\sin x > \cos x$ in the interval $45° < x < 225°$.

5. A triangle has sides of 5 and 7. What is the possible range of values for the third side?

Solution   Since the sum of two sides must be greater than the third side, the third side must be less than 12.

Call the third side x. Then, by the same principle,

$$5 + x > 7 \quad \text{or} \quad x > 2$$
$$2 < x < 12$$

6. In $\triangle KLM$, $\angle K = 60°$ and $\angle M = 50°$. Which side of the triangle is the longest?

Solution   The longest side lies opposite the largest angle. Since the sum of two angles is $110°$, the third angle, L, must be $70°$. The longest side must lie opposite $\angle L$, which is KM.

7. Find the solution set of $2x^2 - x - 3 < 0$, where x is a real number.

Solution $(2x - 3)(x + 1) < 0$

Either $2x - 3 < 0$ and $x + 1 > 0$ or $2x - 3 > 0$ and $x + 1 < 0$

$x < \frac{3}{2}$ and $x > -1$ or $x > \frac{3}{2}$ and $x < -1$, which is impossible.

$$-1 < x < \frac{3}{2}$$

8. What are all p such that $\dfrac{p+1}{p} \leq 1$?

(A) $p > 0$ (B) $p < 0$ (C) $p \leq 0$ (D) $-1 < p < 0$ (E) $-1 \leq p < 0$

Solution▶ (B) $\dfrac{p+1}{p} \leq 1$, p cannot equal zero.

$$1 + \dfrac{1}{p} \leq 1$$
$$\dfrac{1}{p} \leq 0$$
$$p < 0$$

9. If $\log x \geq \log 2 + \dfrac{1}{2} \log x$, then

(A) $x \geq 2$ (B) $x \leq 2$ (C) $x \leq 4$ (D) $x \geq 4$ (E) $x \geq 1$

Solution▶ (D)
$$\log x \geq \log 2 + \log x^{1/2}$$
$$\log x - \log x^{1/2} \geq \log 2$$
$$\log \dfrac{x}{x^{1/2}} \geq \log 2$$
$$x^{1/2} \geq 2$$
$$x \geq 4$$

9. Verbal Problems

When solving verbal problems, follow the steps below:

1. Read the problem carefully and determine the nature of the problem.

2. Consider the given information and data and what is to be found. Represent algebraically the unknown quantity or quantities.

3. Study the relationships of the data in the problem. Draw a diagram, if applicable (motion problems, geometry problems, mixture problems, etc.).

4. Formulate the equation or equations using the representation assigned to the unknown quantities.

5. Solve the equation or equations.

6. Check the results in the original problem.

ILLUSTRATIVE PROBLEMS

1. The area of a rectangular plot is 204 sq ft and its perimeter is 58 ft. Find its dimensions.

Solution▸   Let x = length, y = width.

$$xy = 204$$
$$2x + 2y = 58$$
$$x + y = 29$$
$$y = 29 - x$$

Substitute the value in the area equation.

$$x(29 - x) = 204$$
$$29x - x^2 = 204$$
$$x^2 - 29x + 204 = 0$$
$$(x - 12)(x - 17) = 0$$

$x - 12 = 0$	$x - 17 = 0$
$x = 12$	$x = 17$
$y = 17$	$y = 12$

The length is 17 and the width is 12.

2. Ten lbs of a salt water solution is 20% salt. How much water must be evaporated to strengthen it to a 25% solution?

Solution▸

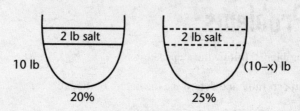

Solution▸   Let x = lb of water evaporated.

$$2 = .25(10 - x) = \frac{1}{4}(10 - x)$$
$$8 = 10 - x$$
$$x = 10 - 8$$
$$= 2 \text{ lb}$$

9. VERBAL PROBLEMS   41

3. A man walked into the country at the rate of 3 mph and hurried back over the same road at 4 mph. The round trip took $5\frac{1}{4}$ hours. How far into the country did he walk?

Solution▸

	rate	time
⟶	3 mph	$\frac{d}{3}$
d miles ⟵	4 mph	$\frac{d}{4}$

$$\frac{d}{3}+\frac{d}{4}=5\frac{1}{4}=\frac{21}{4}$$

Multiply both sides by 12.

$$4d + 3d = 63$$
$$7d = 63$$
$$d = 9$$

4. If the price of an item drops 10 cents per dozen, it becomes possible to buy 2 dozen more items for $6.00 than was possible at the original price. Find the original price.

Solution▸   Let p = original price in cents per dozen
  n = the original number of dozen bought for $6.00

$$pn = 600$$
$$(p - 10)(n + 2) = 600$$
$$pn - 10n + 2p - 20 = 600$$

Substitute $pn = 600$ and $n = \frac{600}{p}$ in second equation.

$$\cancel{600} - \frac{6000}{p} + 2p - 20 = \cancel{600}$$

Multiply through by p.

$$2p^2 - 20p - 6000 = 0$$
$$p^2 - 10p - 3000 = 0$$
$$(p - 60)(p + 50) = 0$$
$$p = 60 \text{ or } p = -50 \text{ (reject)}$$
$$p = 60¢ \text{ per dozen}$$

5. Two planes start from the same place at the same time. One travels east at r mph and the other north at s mph. How far apart will they be after t hours?

Solution▸

Right triangle with legs st and rt, hypotenuse x.

$$x^2 = (rt)^2 + (st)^2$$
$$x^2 = r^2t^2 + s^2t^2$$
$$x^2 = t^2(r^2 + s^2)$$
$$x = t\sqrt{r^2 + s^2}$$

6. The sum of the digits of a two-digit number is 9. If the digits are reversed, the resulting number exceeds the original number by 27. What is the original number?

Solution▶ Let t = ten's digit and u = unit's digit.

$$10t + u = \text{original number}$$
$$10u + t = \text{reversed number}$$
$$t + u = 9$$
$$10u + t - (10t + u) = 27$$
$$9u - 9t = 27$$
$$u - t = 3$$
$$\underline{u + t = 9}$$
$$2u = 12$$
$$u = 6$$
$$t = 3$$

The original number is 36.

10. Geometry

The following formulas and relationships are important in solving geometry problems.

Angle Relationships

1. The *base angles* of an *isosceles* triangle are equal.

2. The sum of the *interior* angles of any n-sided polygon is $180(n - 2)$ degrees.

3. The sum of the *exterior* angles of any n-sided polygon is 360°.

4. If two *parallel* lines are cut by a transversal, the *alternate interior* angles are equal, and the *corresponding* angles are equal.

Angle Measurement Theorems

1. A *central* angle of a circle is measured by its intercepted arc.

2. An *inscribed* angle in a circle is measured by one-half its intercepted arc.

3. An angle formed by *two chords* intersecting *within* a circle is measured by one-half the sum of the opposite intercepted arcs.

4. An angle formed by a *tangent* and a *chord* is measured by one-half its intercepted arc.

5. An angle formed by two *secants*, or by two *tangents,* or by a *tangent* and a *secant,* is measured by one-half the *difference* of the intercepted arcs.

Proportion Relationships

1. A line *parallel* to one side of a triangle divides the other two sides *proportionally*.

2. In two similar triangles, *corresponding* sides, medians, altitudes or angle bisectors are *proportional*.

3. If two *chords* intersect within a circle, the product of the segments of one is equal to the product of the segments of the other.

4. If a *tangent* and a *secant* are drawn to a circle from an outside point, the tangent is the *mean proportional* between the secant and the external segment.

5. In *similar polygons* the *perimeters* have the same ratio as any pair of corresponding sides.

Right Triangle Relationships

1. If an *altitude* is drawn to the hypotenuse of a right triangle, it is the *mean proportional* between the segments of the hypotenuse, and either leg is the mean proportional between the hypotenuse and the segment adjacent to that leg.

2. In a right triangle, the square of the hypotenuse is equal to the sum of the squares of the legs. (Remember the Pythagorean triples: 3, 4, 5; 5, 12, 13.)

3. In a 30°–60° right triangle, the leg opposite the 30° angle is one-half the hypotenuse, and the leg opposite the 60° angle is one-half the hypotenuse times $\sqrt{3}$.

4. In a right isosceles triangle either leg is equal to the hypotenuse times $\sqrt{2}$.

5. In an equilateral triangle of side s, the altitude equals $\frac{s}{2}\sqrt{3}$.

Area Formulas

1. Area of a rectangle = bh (b = base, h = altitude)

2. Area of a parallelogram = bh

3. Area of a triangle = $\frac{1}{2}bh$

4. Area of an equilateral triangle of side $s = \frac{s^2}{4}\sqrt{3}$

5. Area of a trapezoid = $\frac{1}{2}h(b+b')$ where h = altitude and b and b' are the two bases

6. Area of a rhombus = $\frac{1}{2}$ the product of the diagonals

7. Area of a regular polygon = $\frac{1}{2}ap$ where a = apothem and p = perimeter

8. The areas of two similar polygons are to each other as the squares of any two corresponding sides.

Circle Formulas

1. The *circumference* C of a circle of radius r is given by the formula $C = 2\pi r$.

2. The *area* A of a circle of radius r is given by the formula $A = \pi r^2$.

3. The *areas* of two circles are to each other as the *squares* of their radii.

4. The *length* L of an *arc* of $n°$ in a circle of radius r is given by the formula
$L = \frac{n}{360} \times 2\pi r$.

5. The *area* A of a *sector* of a circle of radius r with central angle of $n°$ is given by $A = \frac{n}{360} \times \pi r^2$.

6. The *area* of a *segment* of a circle whose arc is $n°$ is equal to the area of the *sector* of $n°$ minus the area of the isosceles *triangle* with vertex angle of $n°$.

Volume Formulas

1. The *volume* of a *cube* is equal to the cube of an edge.

2. The *volume* of a rectangular *solid* is the product of the length, width, and height.

3. The *volume* V of a right, circular *cylinder* of radius r and height h is given by the formula $V = \pi r^2 h$. The lateral surface area L of such a cylinder is given by the formula $L = 2\pi rh$. The total surface area T is given by the formula $T = 2\pi rh + 2\pi r^2$.

4. The *volume* of a *sphere* of radius r is given by the formula $V = \frac{4}{3}\pi r^3$. The surface area S of the sphere is given by the formula $S = 4\pi r^3$.

5. The volume of a right circular cone of radius r and altitude h is given by the formula $V = \frac{1}{3}\pi r^2 h$.

ANGLE RELATIONSHIPS

1. In $\triangle RST$, if $RS = ST$ and angle $T = 70°$, what is the value, in degrees, of angle S?

Solution

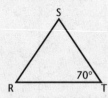

$RS = ST, \angle R = \angle T = 70°$
$R + S + T = 180°$
$70° + S + 70° = 180°$
$\angle S = 40°$

2. In right triangle PQR, RH and RM are altitude and median to the hypotenuse. If angle $Q = 32°$, find angle HRM.

Solution

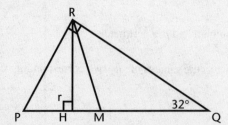

Since $\angle PRQ = 90°$ and $\angle Q = 32°$,
$\angle P = 180° - 90° - 32° = 58°$;
$\angle PRH = 90° - 58° = 32°$.
In $\triangle RMQ$, $RM = MQ$, since the median to the hypotenuse $= \frac{1}{2}$ the hypotenuse. Thus, $\angle MRQ = 32°$.
$\angle HRM = 90° - 32° - 32° = 26°$

10. GEOMETRY   45

3. In the figure, QS and RS are angle bisectors. If ∡P = 80°, how many degrees in ∡QSR?

Solution▸

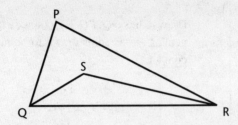

$$\angle Q + \angle R = 180° - P = 180 - 80$$
$$= 100°$$
$$\angle SQR + \angle SRQ = \frac{1}{2} \cdot 100°$$
$$= 50°$$
$$\angle QSR = 180 - 50 = 130°$$

4. How many sides does a regular polygon have if each interior angle equals 176°?

Solution▸

Each exterior angle = 180° − 176° = 4°
Since the sum of the exterior angles is 360°, the number of exterior angles $= \frac{360}{4} = 90$.
The polygon has 90 sides.

5. In the figure, PQRS is a square and RST is an equilateral triangle. Find the value of x.

Solution▸

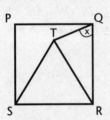

Since ∡TRS = 60° and ∡SRQ = 90°,
∡TRQ = 30°.
RT = SR and QR = SR so that RT = QR, sides of an isosceles △, and ∡RTQ = x°.
Since ∡RTQ + x = 180° − 30° = 150°,
x must equal $\frac{1}{2} \cdot 150° = 75°$.

RIGHT TRIANGLE RELATIONSHIPS

1. A ladder 10 ft tall is standing vertically against a wall that is perpendicular to the ground. The foot of the ladder is moved along the ground 6 ft away from the wall. How many ft down the wall does the top of the ladder move?

Solution▸

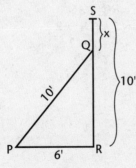

x = SQ, RS = 10
Since PQ = 10 and PR = 6, QR² =
10² − 6² = 100 − 36 = 64
QR = 8
x = 10 − 8 = 2

2. A boat travels 40 m east, 80 m south, then 20 m east again. How far is it from the starting point?

Solution▸

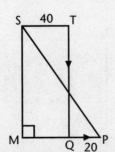

In the figure, draw SM ∥ TQ.
Then SM = TQ = 80 and MQ = ST = 40.
In right △PMS, MP = 60 and SM = 80.
By the Pythagorean Theorem, it follows that SP = 100.

3. Find the length in inches of a tangent drawn to a circle of 8 in. radius from a point 17 in. from the center of the circle.

Solution

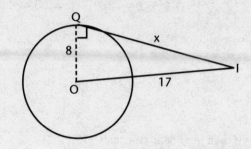

Draw radius OQ; $PQ \perp OQ$ since a tangent $\perp$ to a radius drawn to point of contact.

$$8^2 + x^2 = 17^2$$
$$64 + x^2 = 289$$
$$x^2 = 225$$
$$x = 15$$

4. In the figure, $PQ = PR$, $MS \perp PQ$ and $MT \perp PR$. If $MS = 5$ and $MT = 7$, find the altitude QH.

Solution

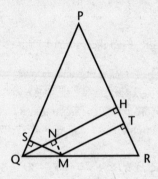

Draw $MN \perp QH$.
Since $MNHT$ is a rectangle, it follows that $NH = MT = 7$. Since $\triangle QMS$ is congruent to $\triangle QMN$, $QN = MS = 5$.
$$QH = QN + NH = 5 + 7 = 12$$

5. Given triangle ABC, $AC \perp BC$, $AD = DB$, $DC = BC$. If $BC = 1$, what is the length of AB?

(A) $\sqrt{10}$ (B) 2   (C) $\sqrt{4 + 2\sqrt{2}}$ (D) $4 + 2\sqrt{2}$ (E) $4\sqrt{2}$

Solution

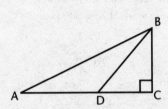

(C) $DC = BC = 1$, so that $DB = \sqrt{2}$.
$AD = DB = \sqrt{2}$
and $AC = AD + DC = 1 + \sqrt{2}$
In right $\triangle ABC$,
$AB^2 = BC^2 + AC^2$
$= 1 + (1 + \sqrt{2})^2$
$= 1 + 1 + 2\sqrt{2} + 2$
$= 4 + 2\sqrt{2}$
$AB = \sqrt{4 + 2\sqrt{2}}$

6. A regular octagon is formed by cutting off each corner of a square whose side is 8. Find the length of one side.

Solution

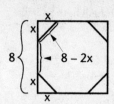

From the figure we see that

$$(8-2x)^2 = x^2 + x^2$$
$$64 - 32x + 4x^2 = 2x^2$$
$$2x^2 - 32x + 64 = 0$$
$$x^2 - 16x + 32 = 0$$
$$x = \frac{16 \pm \sqrt{16^2 - 4(32)}}{2} = \frac{16 \pm 8\sqrt{2}}{2}$$

$= 8 - 4\sqrt{2}$ (because the + sign gives result > 8, which is impossible)

$$\text{side} = 8 - 2x = 8 - 16 + 8\sqrt{2}$$
$$= 8\sqrt{2} - 8$$

7. If the centers of two intersecting circles are 10 in. apart and if the radii of the circles are 6 in. and 10 in. respectively, what is the length of their common chord, in inches?

Solution

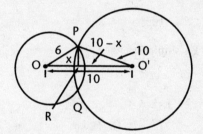

From the figure, $PQ \perp OO'$

Let $y = PR = \dfrac{PQ}{2} = \dfrac{h}{2}; h = 2y$

$$y^2 = 6^2 - x^2 = 10^2 - (10-x)^2$$
$$36 - x^2 = 100 - 100 + 20x - x^2$$
$$x = \frac{9}{5}$$
$$y^2 = 6^2 - \left(\frac{9}{5}\right)^2$$
$$= 36 - \frac{81}{25} = \frac{900 - 81}{25} = \frac{819}{25}$$
$$y = \frac{1}{5}\sqrt{9 \cdot 91} = \frac{3}{5}\sqrt{91}$$

PROPORTION RELATIONSHIPS

1. Two circles of radii 3 in. and 6 in. have their centers 15 in. apart. Find the length in inches of the common internal tangent.

Solution

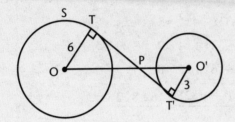

$OT \perp TT'$ and $O'T' \perp TT'$. Thus,
$\triangle OTP \sim \triangle O'T'P'$.

$$\frac{OP}{O'P} = \frac{6}{3} = \frac{2}{1}$$

$$OP = 2O'P$$

$$OP + O'P = 15$$

$$2O'P + O'P = 15$$

$$3O'P = 15$$

$$O'P = 5, \quad OP = 10$$

In right $\triangle OPT$, $PT = 8$. In right $\triangle O'PT'$, $PT' = 4$. Thus, $TT' = 12$

2. One side of a given triangle is 18 in. Inside the triangle a line segment is drawn parallel to this side cutting off a triangle whose area is two-thirds that of the given triangle. Find the length of this segment in inches.

   (A) 12   (B) $6\sqrt{6}$ (C) $9\sqrt{2}$ (D) $6\sqrt{3}$ (E) 9

Solution

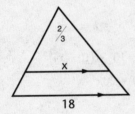

(B) By similar triangles

$$\frac{x^2}{18^2} = \frac{2}{3}$$

$$3x^2 = 2 \cdot 18 \cdot 18$$

$$x^2 = 2 \cdot 6 \cdot 6 \cdot 3$$

$$x = 6\sqrt{6}$$

3. In the figure, $MN \parallel BC$ and MN bisects the area of $\triangle ABC$. If $AD = 10$, find ED.

Solution

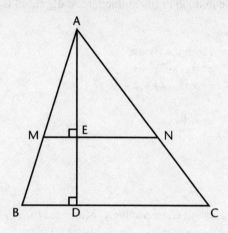

Let $ED = x$
$AE = 10 - x$

Since $\triangle AMN \sim \triangle ABC$, it follows that

$$\frac{\triangle AMN}{\triangle ABC} = \frac{(10-x)^2}{10^2} = \frac{1}{2}$$

Cross multiply.

$$2(10-x)^2 = 100$$
$$(10-x)^2 = 50$$
$$100 - 20x + x^2 - 50 = 0$$
$$x^2 - 20x + 50 = 0$$

$$x = \frac{20 \pm \sqrt{400 - 200}}{2}$$

$$= \frac{20 \pm 10\sqrt{2}}{2} = 10 \pm 5\sqrt{2}$$

Reject positive value since $x < 10$.
$$ED = 10 - 5\sqrt{2}$$

4. In circle O, RS is a diameter and PR is a tangent. If $PQ = 9$ and $QS = 16$, find RS.

Solution

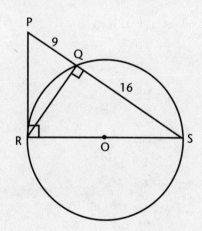

$PR \perp RS$ since a tangent $\perp$ a radius drawn to point of contact. $RQ \perp QS$ since an angle inscribed in a semi-circle is a right angle, and $\angle S$ is common to $\triangle QRS$ and $\triangle PRS$. Thus $\triangle QRS \sim \triangle PRS$. Corresponding sides are proportional.

$$\frac{16}{RS} = \frac{RS}{25}$$
$$RS^2 = (16)(25)$$
$$RS = 4 \cdot 5 = 20$$

CIRCLES

1. If a chord 12 in. long is drawn in a circle and the midpoint of the minor arc of the chord is 2 in. from the chord, what is the radius of the circle?

Solution▸

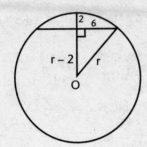

From the figure,
$r^2 = (r-2)^2 + 6^2$
$r^2 = r^2 - 4r + 4 + 36$
$4r = 40$
$r = 10$

2. Lines AB and AC are tangents to a circle at points B and C, respectively. Minor arc BC is 7π in. and the radius of the circle is 18 in. What is the number of degrees in angle BAC?

(A) 90   (B) 95   (C) 70   (D) 100   (E) 110

Solution▸

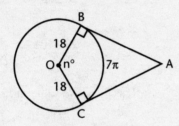

(E) Let $\angle BOC = n°$.

then $7\pi = \dfrac{n}{360} \cdot 2\pi \cdot 18$

$7\pi = \dfrac{n\pi}{10}$

$n = 70°$

$\angle BAC = 180° - 70 = 110°$

3. A circle passes through one vertex of an equilateral triangle and is tangent to the opposite side at its midpoint. What is the ratio of the segments into which the circle divides one of the other sides?

Solution▸

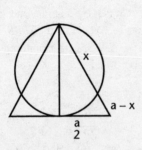

Let side of $\triangle = a$. Then

$a(a-x) = \left(\dfrac{a}{2}\right)^2$

$a^2 - ax = \dfrac{a^2}{4}$

$\dfrac{3a^2}{4} = ax$

$x = \dfrac{3}{4}a$

$a - x = \dfrac{1}{4}a$

$\dfrac{x}{a-x} = \dfrac{3}{1}$

10. GEOMETRY   51

4. Regular pentagon *PQRST* is inscribed in circle *O*. If diagonals *PR* and *QT* intersect in *M*, find the number of degrees in angle *PMQ*.

Solution

Each arc of the circle = $\frac{1}{5} \cdot 360° = 72°$

$\angle PMQ = \frac{1}{2}(\widehat{PQ} + \widehat{TR})$

$= \frac{1}{2}(72° + 144°)$

$= \frac{1}{2}(216°)$

$= 108°$

5. Two tangents are drawn to a circle from a point outside. If one of the intercepted arcs is 140°, how many degrees are in the angle formed by the two tangents?

Solution

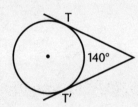

The major arc $TT' = 360° - 140° = 220°$

$\angle P = \frac{1}{2}(220° - 140°)$

$= \frac{1}{2}(80°) = 40°$

6. From the extremities of diameter *PQ* of circle *O*, chords *PR* and *QS* are drawn, intersecting within the circle at *T*. If arc *RS* is 50°, how many degrees in angle *STR*?

Solution

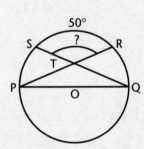

$\angle STR = \frac{1}{2}(\widehat{SR} + \widehat{PQ})$

$= \frac{1}{2}(50° + 180°)$

$= \frac{1}{2}(230°)$

$= 115°$

AREA

1. If circle *R* of area 4 sq in. passes through the center of, and is tangent to, circle *S*, then the area of circle *S*, in square inches, is

   (A) 8   (B) $8\sqrt{2}$ (C) $16\sqrt{2}$ (D) 12   (E) 16

Solution

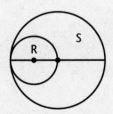

(E) *R* is internally tangent to *S* and its diameter is half that of *S*. Hence *S* has an area 4 times that of *r*, or 16 sq in.

2. Five equal squares are placed side by side to make a single rectangle whose perimeter is 240 in. Find the area of one of these squares in square inches.

Solution

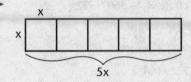

perimeter = $2(5x + x) = 12x = 240$ so that $x = 20$
Area = $x^2 = 20^2 = 400$

3. An altitude h of a triangle is twice the base to which it is drawn. If the area of the triangle is 169 sq cm, how many centimeters in the altitude?

Solution

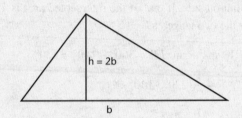

Area of $\triangle = \frac{1}{2}bh = \frac{1}{2} \cdot b \cdot 2b = 169$
$b^2 = 169$
$b = 13$
$h = 2b = 26$ cm

4. Point O is the center of both circles. If the area of shaded region is 9π and $OB = 5$, find the radius of the unshaded region.

Solution

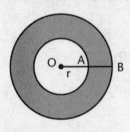

Let $OA = r$. Then $25\pi - \pi r^2 = 9\pi$
$25 - r^2 = 9$
$r^2 = 16$
$r = 4$

5. Given rectangle $ABCD$, semicircles O and P with diameters of 8. If $CD = 25$, what is the area of the shaded region?

Solution

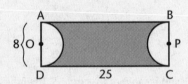

Area of rectangle = $25 \cdot 8 = 200$
Area of both semicircles = $\pi r^2 = 16\pi$
Area of shaded region = $200 - 16\pi$

10. GEOMETRY

6. Find the ratio of the area of a circle inscribed in a square to the area of the circle circumscribed about the square.

Solution

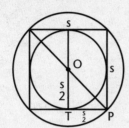

Let the side of the square be s. Then the radius of the inscribed circle is $\frac{s}{2}$.

Since $\triangle OTP$ is right isosceles, the radius OP of the circumscribed circle is

$\frac{s}{2}\sqrt{2}$.

Area of inner circle = $\pi \left(\frac{s}{2}\right)^2 = \frac{\pi s^2}{4}$

Area of outer circle =

$\pi \left(\frac{s}{2}\right)^2 \cdot 2 = \frac{\pi s^2}{4} \cdot 2 = \frac{\pi s^2}{2}$

Ratio = $\dfrac{\frac{\pi s^2}{4}}{\frac{\pi s^2}{2}} = \dfrac{\frac{1}{4}}{\frac{1}{2}} = \frac{2}{4} = \frac{1}{2}$

7. A square is inscribed in a circle. What is the ratio of the area of the square to that of the circle?

Solution

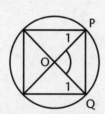

In the figure let $OP = OQ = 1$

In right $\triangle OPQ$, $PQ = \sqrt{2}$

Area of square = $\overline{PQ}^2 = \left(\sqrt{2}\right)^2 = 2$

Area of circle = $\pi (1)^2 = \pi$

Ratio = $\frac{2}{\pi}$

SOLID FIGURES

1. If the radius of a right circular cylinder is tripled, what must be done to the altitude to keep the volume the same?

Solution

$$V = \pi r^2 h$$

Tripling r has the effect of multiplying V by 9. To keep the volume constant, h has to be divided by 9.

2. The surface area of a sphere is 100 sq in. What is the surface area of a sphere having a radius twice the radius of the given sphere?

Solution Since $S = 4\pi r^2 = 100$,

$$\frac{S'}{S} = \left(\frac{r'}{r}\right)^2$$

$$\frac{S'}{100} = \frac{(r')^2}{r^2} = \left(\frac{2}{1}\right)^2 = \frac{4}{1}$$

$$S' = 400$$

54 SAT II: Math

3. The ratio of the diagonal of a cube to the diagonal of a face of the cube is

   (A) $2:\sqrt{3}$ (B) $3:\sqrt{6}$ (C) $3:\sqrt{2}$ (D) $\sqrt{3}:1$ (E) $\sqrt{6}:3$

Solution▶   (B) Let each side of cube equal 1.

$$\text{diagonal of cube: } D = \sqrt{1^2 + 1^2 + 1^2} = \sqrt{3}$$

$$\text{diagonal of face: } D' = \sqrt{1^2 + 1^2} = \sqrt{2}$$

$$\frac{D}{D'} = \frac{\sqrt{3}}{\sqrt{2}} \cdot \frac{\sqrt{3}}{\sqrt{3}} = \frac{3}{\sqrt{6}}$$

4. A pyramid is cut by a plane parallel to its base at a distance from the base equal to two-thirds the length of the altitude. The area of the base is 18. Find the area of the section determined by the pyramid and the cutting plane.

   (A) 1   (B) 2   (C) 3   (D) 6   (E) 9

Solution▶   (B) Let the area of the section be A.

$$\frac{A}{18} = \left(\frac{1}{3}\right)^2 = \frac{1}{9}$$

$$A = 2$$

5. Two spheres of radius 8 and 2 are resting on a plane table top so that they touch each other. How far apart are their points of contact with the plane table top?

   (A) 6   (B) 7   (C) 8   (D) $8\sqrt{2}$ (E) 9

Solution▶   (C)

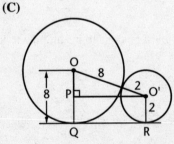

$$OO' = 8 + 2 = 10$$
$$OP = 8 - 2 = 6$$
$$O'P = 8$$
$$QR = 8$$

6. If the radius of a sphere is doubled, the percent increase in volume is

   (A) 100   (B) 200   (C) 400   (D) 700   (E) 800

Solution▶   (D) Let the original radius = 1.

$$\text{The original volume} = \frac{4}{3}\pi \cdot 1^3 = \frac{4}{3}\pi$$

The new radius = 2.

$$\text{The new volume} = \frac{4}{3}\pi \cdot 2^3 = \frac{4}{3}\pi \cdot 8$$

$$\text{The increase in volume} = \frac{4}{3}\pi(8-1) = \frac{4}{3}\pi \cdot 7$$

$$\text{The percent increase} = \frac{\frac{4}{3}\pi \cdot 7}{\frac{4}{3}\pi} \times 100$$

$$= 700$$

7. A right circular cylinder is circumscribed about a sphere. If S represents the surface area of the sphere and T represents the total area of the cylinder, then

(A) $S = \dfrac{2}{3}T$ (B) $S < \dfrac{2}{3}T$ (C) $S > \dfrac{2}{3}T$ (D) $S \le \dfrac{2}{3}T$ (E) $S \ge \dfrac{2}{3}T$

Solution▸ (A)

$S = 4\pi r^2$
$T = 2\pi r^2 + 2\pi r(2r) = 6\pi r^2$
$\dfrac{S}{T} = \dfrac{4\pi r^2}{6\pi r^2} = \dfrac{2}{3}$
$S = \dfrac{2}{3}T$

8. A triangle whose sides are 2, 2, and 3 is revolved about its longest side. Find the volume generated. (Use $\pi = \dfrac{22}{7}$)

Solution▸

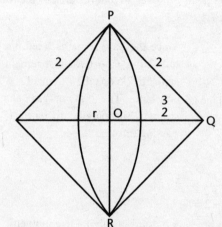

The solid figure formed consists of two congruent cones. In the figure, Q is the apex of one cone and OP the radius r of the base; $h = \left(\dfrac{3}{2}\right)$

$V = \dfrac{1}{3}\pi r^2 h,\ r^2 = 4 - \dfrac{9}{4} = \dfrac{7}{4}$

$V = \dfrac{1}{3} \cdot \dfrac{22}{7} \cdot \dfrac{7}{4} \cdot \dfrac{3}{2}$

$V = \dfrac{11}{4}$

$2V = \dfrac{2}{1} \cdot \dfrac{11}{4} = \dfrac{11}{2} = 5\dfrac{1}{2}$

LOCUS

1. What is the locus of points equidistant from two intersecting lines and at a given distance d from their point of intersection?

Solution▸

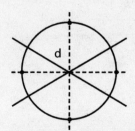

The locus of points equidistant from two intersecting lines consists of the two angle-bisectors of the angles formed by the lines.
At the intersection draw a circle of radius d.
The desired locus is the four points where this circle intersects the angle-bisectors.

2. Two parallel planes are 6 in. apart. Find the locus of points equidistant from the two planes and 4 in. from a point P in one of them.

Solution▸ The locus of points equidistant from the two planes is a parallel plane midway between them (3 in. from each). The locus of points 4 in. from P is a sphere with P as center and radius 4 in. The intersection of this sphere with the mid-plane is a *circle*, the desired locus.

3. Parallel lines r and s are 12 in. apart. Point P lies between r and s at a distance of 5 in. from r. How many points are equidistant from r and s and also 7 in. from P?

Solution▶ All points equidistant from r and s lie on a line parallel to r and s and lying midway between them. All points 7 in. from P lie on a circle of radius 7 and center at P. These two loci intersect at two points.

4. What is the locus of points in space 4 in. from a given plane and 6 in. from a given point in the plane?

Solution▶ The locus of points 4 in. from the given plane consists of two planes parallel to the given plane. The locus of points 6 in. from the given point is a sphere of radius 6. These two loci intersect in two circles.

5. What is the equation of the locus of points equidistant from the points (–2, 5) and (–2, –1)?

Solution▶ The line segment joining the two points is part of the line $x = -2$, a vertical line. The mid-point of the line segment is (–2, 2). The desired locus is the line $y = 2$.

6. Given $\triangle PQR$. The base QR remains fixed and the point P moves so that the area of the triangle PQR remains constant. What is the locus of point P?

Solution▶ 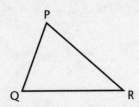 Since the base remains fixed, the altitude from P to QR must remain constant to keep the area of $\triangle PQR$ constant. Thus P moves along a straight line parallel to base QR and passing through P.

11. Trigonometry

The following trigonometric formulas and relationships are very helpful in solving trigonometric problems.

Relationships Among the Functions

1. Reciprocal and Quotient Relationships

$$\csc A = \frac{1}{\sin A}$$
$$\sec A = \frac{1}{\cos A}$$
$$\cot A = \frac{1}{\tan A}$$
$$\tan A = \frac{\sin A}{\cos A}$$
$$\cot A = \frac{\cos A}{\sin A}$$

2. Pythagorean Relationships

$\sin^2 A + \cos^2 A = 1$
$\sec^2 A = 1 + \tan^2 A$
$\csc^2 A = 1 + \cot^2 A$

3. The trigonometric function of any angle A is equal to the co-function of the complementary angle $(90 - A)$. Thus, $\sin A = \cos(90 - A)$, etc.

Functions of the Sum of Two Angles

$\sin(A + B) = \sin A \cos B + \cos A \sin B$
$\cos(A + B) = \cos A \cos B - \sin A \sin B$
$\tan(A + B) = \dfrac{\tan A + \tan B}{1 - \tan A \tan B}$

Double Angle Formulas

$\sin 2x = 2 \sin x \cos x$
$\cos 2x = \cos^2 x - \sin^2 x = 1 - 2\sin^2 x$
$\tan 2x = \dfrac{2 \tan x}{1 - \tan^2 x}$

Half Angle Formulas

$\sin \dfrac{1}{2} A = \pm \sqrt{\dfrac{1 - \cos A}{2}}$

$\cos \dfrac{1}{2} A = \pm \sqrt{\dfrac{1 + \cos A}{2}}$

$\tan \dfrac{1}{2} A = \pm \sqrt{\dfrac{1 - \cos A}{1 + \cos A}}$

Relationships of Sides to Angles in a Triangle

Law of Sines: $\dfrac{a}{\sin A} = \dfrac{b}{\sin B} = \dfrac{c}{\sin C}$

Law of Cosines: $c^2 = a^2 + b^2 - 2ab \cos C$

Area Formulas for a Triangle

$K = \dfrac{1}{2} ab \sin C$

$K = \sqrt{s(s-a)(s-b)(s-c)}$ where $s = \dfrac{a+b+c}{2}$

58 SAT II: Math

Graphs of Trigonometric Functions

1. If the equation of a curve is of the form $y = b \sin nx$ or $y = b \cos nx$, $n > 0$, the *amplitude* of the curve $= b$, the *period* of the curve $= \frac{360°}{n}$ or $\frac{2\pi}{n}$ radians, and the *frequency* of the curve is the number of cycles in 360° or 2π radians, which equals n.

2. If the equation of a curve is of the form $y = b \tan nx$ or $y = b \cot nx$, $n > 0$, the *period* of the curve $= \frac{180°}{n}$ or $\frac{\pi}{n}$ radians, and the *frequency* of the curve is the number of cycles in 180° or π radians, which equals n.

DEGREE AND RADIAN MEASURE

1. Expressed in radians, an angle of 108° is equal to

   (A) $\frac{2\pi}{3}$ (B) $\frac{6\pi}{5}$ (C) $\frac{3\pi}{10}$ (D) $\frac{2\pi}{5}$ (E) $\frac{3\pi}{5}$

Solution▶ (E)

$$\frac{108}{x} = \frac{180}{\pi}$$
$$180x = 108\pi$$
$$x = \frac{108\pi}{180} = \frac{9\pi}{15} = \frac{3\pi}{5}$$

2. The radius of a circle is 9 in. Find the number of radians in a central angle which subtends an arc of 1 ft in this circle.

Solution▶

$$l = r\theta$$
$$12 = 9\theta$$
$$\theta = \frac{12}{9} = \frac{4}{3} \text{ radians}$$

3. The value of $\cos \frac{2\pi}{3}$ is

   (A) $-\frac{1}{2}$ (B) $\frac{1}{2}$ (C) $-\frac{\sqrt{3}}{2}$ (D) $\frac{\sqrt{3}}{2}$ (E) $\frac{\sqrt{2}}{2}$

Solution▶ (A)

$$\cos \frac{2\pi}{3} = -\cos \frac{\pi}{3} = -\frac{1}{2}$$

4. If, in a circle of radius 5, an arc subtends a central angle of 2.4 radians, the length of the arc is

   (A) 24 (B) .48 (C) 3π (D) 5π (E) 12

Solution▶ (E)

$$l = r\theta$$
$$l = 5(2.4)$$
$$= 12$$

5. The bottom of a pendulum describes an arc 3 ft long when the pendulum swings through an angle of $\frac{1}{2}$ radian. The length of the pendulum in ft is

(A) 2   (B) 3   (C) 4   (D) 5   (E) 6

Solution▶   (E)

$$l = r\theta$$
$$3 = \frac{1}{2}r$$
$$r = 6$$

TRIGONOMETRIC IDENTITIES

1. Express the function $\sin x$ in terms of $\tan x$.

Solution▶ $\sin x = \tan x \cos x = \tan x \sqrt{1 - \sin^2 x}$
Square both sides.

$$\sin^2 x = \tan^2 x (1 - \sin^2 x)$$
$$= \tan^2 x - \tan^2 x \sin^2 x$$
$$\sin^2 x + \sin^2 x \tan^2 x = \tan^2 x$$

Factor the left member.

$$\sin^2 x \left(1 + \tan^2 x\right) = \tan^2 x$$

$$\sin^2 x = \frac{\tan^2 x}{1 + \tan^2 x}$$

Take the square root of both sides.

$$\sin x = \pm \frac{\tan x}{\sqrt{1 + \tan^2 x}}$$

2. If $\log \tan x = k$, express $\cot x$ in terms of k.

Solution▶

$$\log \cot x = \log \frac{1}{\tan x}$$
$$= \log 1 - \log \tan x$$
$$= 0 - k$$
$$= -k$$

3. Express $(1 + \sec y)(1 - \cos y)$ as a product of two trigonometric functions of y.

Solution▶   Multiply the two binomials.

$$1 + \sec y - \cos y - \sec y \cos y$$

Substitute $\cos y$ for $\sec y$.

$$1 + \frac{1}{\cos y} - \cos y - \frac{1}{\cos y} \cdot \cos y$$
$$= 1 + \frac{1}{\cos y} - \cos y - 1 = \frac{1}{\cos y} - \cos y$$
$$= \frac{1 - \cos^2 y}{\cos y} = \frac{\sin^2 y}{\cos y} = \frac{\sin y}{\cos y} \cdot \sin y$$
$$= \tan y \cdot \sin y$$

4. Write tan 2x numerically, if $\tan x = \sqrt{2}$.

Solution

$$\tan 2x = \frac{2\tan x}{1-\tan^2 x}$$

$$= \frac{2\sqrt{2}}{1-\left(\sqrt{2}\right)^2} = \frac{2\sqrt{2}}{-1} = -2\sqrt{2}$$

5. Simplify $\dfrac{\sin 2x}{1+\cos 2x}$ and write as a function of x.

Solution

$$\frac{2\sin x \cos x}{1+2\cos^2 x - 1} = \frac{2\sin x \cos x}{2\cos^2 x}$$

$$= \frac{\sin x}{\cos x} = \tan x$$

6. If $\cos 200° = p$, express the value of $\cot 70°$ in terms of p.

Solution

$$\cos 200° = -\cos 20° = p$$
$$\sin 70° = \cos 20° = -p$$
$$\cos 70° = \sqrt{1-\sin^2 70°} = \sqrt{1-p^2}$$
$$\cot 70° = \frac{\cos 70°}{\sin 70°} = \frac{\sqrt{1-p^2}}{-p}$$
$$= -\frac{\sqrt{1-p^2}}{p}$$

LAW OF SINES

1. If in $\triangle ABC$, $A = 30°$ and $B = 20°$, find the ratio $BC : AC$.

Solution Let $BC = a$ and $AC = b$.

$$\frac{BC}{AC} = \frac{a}{b} = \frac{\sin A}{\sin B} = \frac{\sin 30°}{\sin 120°}$$

$$= \frac{\frac{1}{2}}{\frac{\sqrt{3}}{2}} = \frac{1}{\sqrt{3}}$$

2. In triangle ABC, $A = 30°$, $C = 105°$, $a = 8$. Find side b.

Solution If $A = 30°$ and $C = 105°$, then angle $B = 180° - 135° = 45°$.

$$\frac{b}{\sin B} = \frac{a}{\sin A}$$

$$\frac{b}{\sin 45°} = \frac{8}{\sin 30°}$$

$$\frac{b}{\frac{1}{\sqrt{2}}} = \frac{8}{\frac{1}{2}}$$

$$b\sqrt{2} = 16$$

$$b = \frac{16}{\sqrt{2}} \cdot \frac{\sqrt{2}}{\sqrt{2}} = 8\sqrt{2}$$

3. If AB and angles x and y are given, express BD in terms of these quantities.

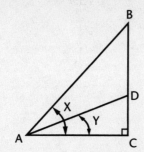

In $\triangle ABD$, by the law of sines,

$$\frac{BD}{\sin(x-y)} = \frac{AB}{\sin \angle BDA} = \frac{AB}{\sin(90°+y)}$$
$$= \frac{AB}{\cos y}$$
$$BD = \frac{AB \sin(x-y)}{\cos y}$$

4. Two sides of a triangle are 5 and 6, and the included angle contains 120°. Find its area.

Solution▶

$$\text{Area of triangle} = \frac{1}{2} ab \sin C$$
$$= \frac{1}{2} \cdot 5 \cdot 6 \cdot \sin 120°$$
$$= 15 \sin 60°$$
$$= \frac{15\sqrt{3}}{2}$$

LAW OF COSINES

1. If the sides of a triangle are 2, 3, and 4, find the cosine of the largest angle.

Solution▶ Let x = the angle opposite the side of 4.
Then, by the law of cosines,

$$4^2 = 2^2 + 3^2 - 2(2)(3) \cos x$$
$$16 = 4 + 9 - 12 \cos x$$
$$3 = -12 \cos x$$
$$\cos x = -\frac{1}{4}$$

2. In $\triangle ABC$, $a = 1$, $b = 1$ and $C = 120°$. Find the value of c.

Solution▶

$$c^2 = a^2 + b^2 - 2ab \cos C$$
$$= 1^2 + 1^2 - 2 \cdot 1 \cdot 1 \cos 120°$$
$$c^2 = 2 - 2(-\cos 60°)$$
$$c^2 = 2 + 2\left(\frac{1}{2}\right) = 3$$
$$c = \sqrt{3}$$

3. In $\triangle ABC$, if $a = 8$, $b = 5$ and $c = 9$, find $\cos A$.

Solution▶

$$a^2 = b^2 + c^2 - 2bc \cos A$$
$$64 = 25 + 81 - 90 \cos A$$
$$64 = 106 - 90 \cos A$$
$$90 \cos A = 42$$
$$\cos A = \frac{42}{90} = \frac{7}{15}$$

GRAPHS OF TRIGONOMETRIC FUNCTIONS

1. How does $\sin x$ change as x varies from $90°$ to $270°$?

Solution▶

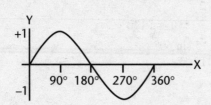

Sketch the graph of $y = \sin x$.
From the graph we see that $\sin x$ decreases continuously from $+1$ to -1.

2. What is the period of the graph of $y = 3 \cos 2x$?

Solution▶ The normal period of $y = \cos x$ is 2π.

For the graph of $y = 3 \cos 2x$, the period $= \frac{2\pi}{2} = \pi$

3. The graph of the function $y = 2 \sin \frac{1}{2} x$ passes through the point whose coordinates are

(A) $(0, 2)$ (B) $\left(\frac{\pi}{2}, 1\right)$ (C) $\left(0, \frac{1}{2}\right)$ (D) $(\pi, 1)$ (E) $(\pi, 2)$

Solution▶   (E) $y = 2 \sin \frac{1}{2} x$

Substitute for x the abscissa of each ordered pair in the five choices. Note that when $x = \pi$,

$$y = 2 \sin \frac{\pi}{2} = 2 \cdot 1 = 2$$

The point $(\pi, 2)$ lies on the graph.

TRIGONOMETRIC EQUATIONS

1. If $\cos x = -\frac{4}{5}$ and $\tan x$ is positive, find the value of $\sin x$.

Solution▶

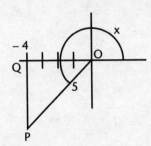

Since $\cos x$ is negative and $\tan x$ is positive, x is in the third quadrant.
In right $\triangle POQ$, the length of OQ is 4 and of OP is 5. It follows that $PO = -3$.
Therefore, $\sin x = -\frac{3}{5}$

2. Find all values of y between $0°$ and $180°$ that satisfy the equation $2\sin^2 y + 3\cos y = 0$.

Solution▶ Substitute $1 - \cos^2 y$ for $\sin^2 y$.

$$2(1 - \cos^2 y) + 3\cos y = 0$$
$$2 - 2\cos^2 y + 3\cos y = 0$$
$$2\cos^2 y - 3\cos y - 2 = 0$$
$$(2\cos y + 1)(\cos y - 2) = 0$$
$$2\cos y + 1 = 0 \quad \text{or} \quad \cos y - 2 = 0$$
$$\cos y = -\frac{1}{2} \quad \text{or} \quad \cos y = 2$$
(Reject $\cos y = 2$ since $\cos y \leq 1$)
$$y = 120°$$

3. How many values of x between $0°$ and $360°$ satisfy the equation $2\sec^2 x + 5\tan x = 0$?

Solution▶
$$2(1 + \tan^2 x) + 5\tan x = 0$$
$$2\tan^2 x + 5\tan x + 2 = 0$$
$$(2\tan x + 1)(\tan x + 2) = 0$$
$$\tan x = -\frac{1}{2}, \tan x = 2$$

For each of the these values of $\tan x$, there are 2 values of x, in Quadrants II and IV and I and III. Hence there are *four* solutions.

4. The value of x between $180°$ and $270°$ which satisfies the equation $\tan x = \cot x$ is

(A) $200°$ (B) $210°$ (C) $225°$ (D) $240°$ (E) $250°$

Solution▶ (C)
$$\tan x = \cot x$$
$$\tan x = \frac{1}{\tan x}$$
$$\tan^2 x = 1$$
$$\tan x = \pm 1$$

In quadrant III, $x = 225°$.

5. Express, in degrees, the measure of the obtuse angle which satisfies the equation $2 \tan \theta \cos \theta - 1 = 0$.

Solution▸   Replace $\tan \theta$ by $\dfrac{\sin \theta}{\cos \theta}$.

$$2\dfrac{\sin \theta}{\cos \theta} \cos \theta - 1 = 0$$
$$2 \sin \theta - 1 = 0$$
$$\sin \theta = \dfrac{1}{2}$$

$\theta = 30°$ or the supplement of $30°$

$\theta = 150°$

12. Graphs and Coordinate Geometry

The following formulas and relationships are important in dealing with problems in coordinate geometry.

1. The distance d between two points whose coordinates are (x_1, y_1) and (x_2, y_2) is

$$d = \sqrt{(x_1 - x_2)^2 + (y_1 - y_2)^2}$$

2. The coordinates of the *midpoint* $M(x, y)$ of the line segment which joins the points (x_1, y_1) and (x_2, y_2) are $x = \dfrac{x_1 + x_2}{2}$ $y = \dfrac{y_1 + y_2}{2}$

3. The *slope m* of a line passing through the points (x_1, y_1) and (x_2, y_2) is given by

$$m = \dfrac{y_1 - y_2}{x_1 - x_2} = \dfrac{y_2 - y_1}{x_2 - x_1}$$

4. If two lines are *parallel*, their slopes are equal; and conversely.

5. If two *perpendicular* lines have slopes m_1 and m_2, then $m_1 = -\dfrac{1}{m_2}$; and conversely.

6. The equation of a line *parallel to the x-axis* is $y = k$ where k is a constant.

7. The equation of a line *parallel to the y-axis* is $x = c$ where c is a constant.

8. The graph of an equation of the form $y = mx + b$ is a *straight line* whose *slope* is m and whose *y*-intercept is b.

9. The equation of a *straight line* passing through a *given point* (x_1, y_1) and having slope m is
$y - y_1 = m(x - x_1)$

10. The graph of the equation $x^2 + y^2 = r^2$ is a *circle* of radius r with center at the origin.

11. The graph of the general *quadratic* function $y = ax^2 + bx + c$ is a *parabola* with an axis of symmetry parallel to the *y*-axis. The equation of the axis of symmetry is $x = -\dfrac{b}{2a}$.

12. Graphs and Coordinate Geometry

12. The graph of $ax^2 + by^2 = c$, where a, b and c are positive, is an *ellipse* with center at the origin. The ellipse is symmetric with respect to the origin.

13. The graph of $ax^2 - by^2 = c$, where a, b and c are positive, is a *hyperbola* symmetric with respect to the origin and having intercepts only on the x-axis.

ILLUSTRATIVE PROBLEMS

1. M is the midpoint of line segment PQ. The coordinates of point P are $(5, -3)$ and of point M are $(5, 7)$. Find the coordinates of point Q.

Solution Let the coordinates of Q be (x, y). Then

$$5 = \frac{1}{2}(5+x) \qquad \qquad 7 = \frac{1}{2}(-3+y)$$
$$10 = 5 + x \quad \text{and} \quad 14 = -3 + y$$
$$x = 5 \qquad \qquad 17 = y$$

Coordinates of Q are $(5, 17)$

2. A triangle has vertices $R(1, 2)$, $S(7, 10)$ and $T(-1, 6)$. What kind of a triangle is RST?

Solution
$$\text{slope of } RS = \frac{10-2}{7-1} = \frac{8}{6} = \frac{4}{3}$$
$$\text{slope of } RT = \frac{6-2}{-1-1} = \frac{4}{-2} = -2$$
$$\text{slope of } ST = \frac{10-6}{7-(-1)} = \frac{4}{8} = \frac{1}{2}$$

Since the slope of ST is the negative reciprocal of the slope of RT, $RT \perp ST$, and the triangle is a right triangle.

3. Find the equation of the straight line through the point $(5, -4)$ and parallel to the line $y = 3x - 2$.

Solution The slope of $y = 3x - 2$ is 3. The desired line, therefore, has slope 3.
By the point-slope form, the equation is

$$y - (-4) = 3(x - 5)$$
$$y + 4 = 3x - 15$$
$$y = 3x - 19$$

4. If the equations $x^2 + y^2 = 16$ and $y = x^2 + 2$ are graphed on the same set of axes, how many points of intersection are there?

Solution 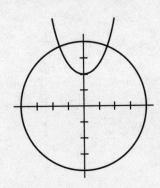 Sketch both graphs as indicated.
$x^2 + y^2$ is a circle of radius 4 and center at origin.
$y = x^2 + 2$ is a parabola. Several points are $(0, 2)$, $(\pm 1, 3)$, $(\pm 2, 6)$.
The graphs intersect in two points.

5. Which of the following points lies inside the circle $x^2 + y^2 = 25$?

   (A) (3, 4)   (B) (–4, 3)   (C) (4, $2\sqrt{2}$)   (D) (4, $2\sqrt{3}$)   (E) none of these

Solution▸ (C) The given circle has a radius of 5 and center at the origin. Points A and B are at distance 5 from the origin, and lie on the circle. The distance of D from the origin is

$$\sqrt{4^2 + (2\sqrt{3})^2} = \sqrt{16 + 12} = \sqrt{28} > 5$$

D lies outside the circle.
The distance of point C from the origin is

$$\sqrt{4^2 + (2\sqrt{2})^2} = \sqrt{16 + 8} = \sqrt{24} < 5$$

C lies inside the circle.

6. For what value of K is the graph of the equation $y = 2x^2 - 3x + K$ tangent to the x-axis?

Solution▸ If this parabola is tangent to the x-axis, the roots of the equation $2x^2 - 3x + K = 0$ are equal, and the discriminant of this equation must be zero.

$$(-3)^2 - 4(2)K = 0$$
$$9 = 8K$$
$$K = \frac{9}{8}$$

7. What is the equation of the graph in the figure?

Solution▸

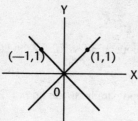

The graph consists of the two straight lines $y = x$ in the first and third quadrants, and $y = -x$ in the second and fourth.
The equation is therefore $|y| = |x|$

8. What is the equation of the locus of points equidistant from the points (3, 0) and (0, 3)?

Solution▸

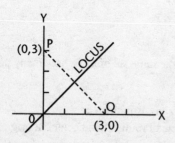

The locus is the perpendicular bisector of the line PQ. This locus is a line bisecting the first quadrant angle.
Its equation is $y = x$.

13. Number Systems and Concepts

The Set of Integers

1. The set of *natural numbers* is made up of the ordinary counting numbers 1, 2, 3, . . .

2. The set of *integers* is made up of all positive and negative whole numbers and zero. An *even* integer is any multiple of two; that is, it is any integer which can be written in the form $2n$, where n is any integer. Thus zero is considered an even integer (2×0). An *odd* integer is any integer which is *not* even.

3. The *sum* of two odd integers or two even integers is an *even* integer. The sum of an odd integer and an even integer is an *odd* integer.

4. The *product* of two odd integers is an *odd* integer. Any power of an odd integer is also an odd integer. The *product* of any integer and an even integer is also an *even* integer.

5. The set of integers is *closed* under addition, subtraction or multiplication; that is, if any of these operations is performed upon two integers, the result will also be an integer. The set of integers is *not* closed under division, since the quotient of two integers is not always an integer.

The Set of Rational Numbers

1. A *rational* number is any number that can be written in the form $\frac{p}{q}$, where p and q are integers and $q \neq 0$. The set of *integers* is a *subset* of the set of *rational* numbers since any integer p can be written in the form of a *ratio* of p to 1.

2. The set of decimal fractions that can be written as *finite* decimals is also a *subset* of the set of rational numbers, since a finite decimal can always be written as the ratio of an integer and a power of 10. We can also show that *infinite* decimals that have *repeating* groups of digits can be expressed as rational numbers.

3. An *irrational* number is any real number that is *not* rational. We can show that numbers like $\sqrt{2}$, $\sqrt[3]{11}$, and π are irrational numbers. Irrational numbers are infinite decimals whose digits do *not* repeat endlessly in groups.

4. The set of rational numbers is *closed* under addition, subtraction, multiplication and division, except for division by zero, which is not defined.

5. The set of *real* numbers is made up of both the set of rational and the set of irrational numbers. It is closed under the four basic operations, except for division by zero.

Properties of Real Numbers

1. The operations of addition and multiplication are *commutative* with respect to the set of real numbers. Thus, if p and q are real numbers

$$p + q = q + p \text{ and } p \cdot q = q \cdot p$$

2. The operations of addition and multiplication are *associative* with respect to the set of real numbers. Thus, if p, q, and r are real numbers, then

$$p + (q + r) = (p + q) + r = p + q + r \text{ and } p \cdot (q \cdot r) = (p \cdot q) \cdot r = pqr$$

3. The multiplication of real numbers is *distributive* over addition. Thus, if p, q, and r are real numbers,

$$p(q + r) = pq + pr$$

4. The number 0 is the *identity* element for addition of real numbers; that is, if p is a real number

$$p + 0 = p$$

5. The number 1 is the *identity* element for multiplication of real numbers; that is, if p is a real number

$$p \cdot 1 = p$$

6. The *additive inverse* of any real number p is $-p$.

$$p + (-p) = 0$$

7. The *multiplicative inverse* of any real number $p (p \neq 0)$ is $\frac{1}{p}$.

$$p\left(\frac{1}{p}\right) = 1$$

We also refer to $\frac{1}{p}$ as the *reciprocal* of p. The reciprocal of 1 is 1.

The Set of Complex Numbers

1. A *complex number* is any number that may be expressed in the form $c + di$ where c and d are real numbers and $i = \sqrt{-1}$. When $c = 0$, the number is called a *pure imaginary* number. When $d = 0$, the number is a real number. The set of *real numbers* is a *subset* of the set of complex numbers.

2. In the complex number $c + di$, c is called the *real* part and d the *imaginary* part of the complex number. Two complex numbers are equal if and only if their real parts are equal and their imaginary parts are equal.

3. Sum of two complex numbers:

$$(a + bi) + (c + di) = (a + c) + (b + d)i$$

Product of two complex numbers:

$$(a + bi)(c + di) = (ac - bd) + (ad + bc)i$$

Thus,

$$i^2 = i \cdot i = (0 + 1i)(0 + 1i)$$
$$= (0 - 1 \cdot 1) + (0 \cdot 1 + 1 \cdot 0)i$$
$$= -1 + 0i = -1$$

13. NUMBER SYSTEMS AND CONCEPTS

4. The set of complex numbers is *closed* under addition, subtraction, multiplication, and division.

5. The number 0 is the *additive identity* element for the set of complex numbers. The number 1 is the *multiplicative identity* element for the set of complex numbers (except for 0).

6. The commutative, associative, and distributive properties apply to the set of complex numbers as they do for the set of real numbers.

ILLUSTRATIVE PROBLEMS

1. Under which arithmetic operations is the set of *even* integers (including zero) closed?

Solution Represent two *even* integers by $2x$ and $2y$ where x and y are integers.
$2x + 2y = 2(x + y)$. Since $(x + y)$ is an integer, $2(x + y)$ is an even integer.
$2x - 2y = 2(x - y)$. Since $(x - y)$ is an integer, $2(x - y)$ is an even integer.
$(2x) \cdot (2y) = 2(2xy)$. Since $2xy$ is an integer, $2(2xy)$ is an even integer.
But $\frac{2x}{2y} = \frac{x}{y}$ need not be an integer.
The even integers (including zero) are closed under addition, subtraction, and multiplication.

2. What complex number (in the form $a + bi$) is the multiplicative inverse of $1 + i$?

Solution
$$(1+i)x = 1$$
$$x = \frac{1}{1+i} \cdot \frac{1-i}{1-i}$$
$$x = \frac{1-i}{1-i^2} = \frac{1-i}{1-(-1)} = \frac{1-i}{2}$$
$$x = \frac{1}{2} - \frac{1}{2}i$$

3. If $\odot$ is an operation on positive real numbers, for which of the following definitions of $\odot$ is $r \odot s = s \odot r$ (commutative property)?

(A) $r \odot s = r - s$
(B) $r \odot s = \frac{r}{s}$
(C) $r \odot s = r^2 s$
(D) $r \odot s = \frac{r+s}{rs}$
(E) $r \odot s = r^2 + rs + s^4$

Solution (D)

(A) is not commutative because $r - s \neq s - r$
(B) is not commutative because $\frac{r}{s} \neq \frac{s}{r}$
(C) is not commutative because $r^2 s \neq s^2 r$
(E) is not commutative because $r^2 + rs + s^4 \neq s^2 + sr + r^4$
(D) *is* commutative because $\frac{r+s}{rs} = \frac{s+r}{sr}$

4. If $a^2 - 2ab + b^2 = m$, where a is an odd and b is an even integer, what kind of an integer is m?

Solution $(a-b)^2 = m$, and so m is a *perfect square*, since $a - b$ is an integer.
Also, the difference between an odd and an even integer is odd, and so $(a - b)$ is odd and $(a - b)^2$ is odd.
m is an odd perfect square.

5. Consider the number 144_b, which is written to the base b, b a positive integer. For what values of b is the number a perfect square?

Solution
$$144_b = 1 \cdot b^2 + 4b + 4$$
$$= (b+2)^2$$

The number is a perfect square for any integral value of b. However, since the digits up to 4 are used to write the number, $b > 4$.

6. Which of the following is an irrational number?

(A) $\dfrac{2}{3}$ (B) $\sqrt[3]{-8}$ (C) $\sqrt{\dfrac{32}{50}}$ (D) $\dfrac{5}{\sqrt{5}}$ (E) none of these

Solution (D)   (A) is a rational fraction.

(B) $\sqrt[3]{-8} = -2$, which is rational $\left(\dfrac{-2}{1}\right)$.

(C) $\sqrt{\dfrac{32}{50}} = \sqrt{\dfrac{16}{25}} = \dfrac{4}{5}$, which is rational.

(D) $\dfrac{5}{\sqrt{5}} = \dfrac{5\sqrt{5}}{\sqrt{5}\sqrt{5}} = \dfrac{5\sqrt{5}}{5} = \sqrt{5}$, which is irrational.

7. $\dfrac{2+i}{2-i} - \dfrac{2-i}{2+i} =$

Solution   Combine the fractions. L.C.D. $= (2-i)(2+i)$

$$\dfrac{(2+i)^2 - (2-i^2)}{(2-i)(2+i)}$$

$$= \dfrac{4 + 4i - 1 - 4 + 4i + 1}{4 - (-1)}$$

$$= \dfrac{8i}{5}$$

8. What is the value of $i^{88} - i^{22}$?

Solution
$$i^{88} - i^{22} = (i^{22})^4 - i^{22}$$
$$i^{22} = i^{20} \cdot i^2 = 1 \cdot i^2 = -1$$
$$i^{88} - i^{22} = (-1)^4 - (-1) = 1 + 1 = 2$$

9. If $f(x) = x^3 + x^2 + 2x + 6$, find $f(i)$.

Solution▶
$$f(i) = i^3 + i^2 + 2i + 6$$
$$= -i - 1 + 2i + 6$$
$$= 5 + i$$

14. Arithmetic and Geometric Progressions

A sequence of numbers such as 4, 7, 10, 13 . . . is called an *arithmetic progression* (A.P.). Note that each term is obtained from the preceding term by adding 3; thus, the difference between any term and its preceding term is 3. We call this number the *common difference* (d) of the progression or sequence. If the successive terms decrease, we consider d to be negative.

If we designate the terms of an A.P. by $a_1, a_2, a_3 \ldots a_n$, we may easily develop the following formula for the nth term, a_n, in terms of a and d:

$$a_n = a_1 + (n-1)d$$

The indicated *sum* of the terms of a progression is called a *series*. $4 + 7 + 10 + 13 + \ldots$ may be referred to as an *arithmetic series* or the sum of an *arithmetic progression*. The sum of the first n terms of an A.P. is given by the formula

$$S_n = \frac{n}{2}(a_1 + a_n)$$

We may convert the S_n formula to a more convenient form by substituting in it $a_n = a_1 + (n-1)d$.

$$S_n = \frac{n}{2}\left[2a_1 + (n-1)d\right]$$

A sequence of terms such as 3, 6, 12, 24 . . . is called a geometric progression (G.P.). Here, the *ratio* (r) of any term to its preceding term is *constant*, in this case, $r = 2$. If we designate the terms of a G.P. by $a_1, a_2, a_3 \ldots a_n$, we can express a_n in terms of a, and r as follows:

$$a_n = a_1 r^{n-1}$$

The sum S_n of n terms of a G.P. is given by the formula:

$$S_n = \frac{a_1 - a_1 r^n}{1 - r}$$

If the absolute value of the ratio, r, of a G.P. is less than 1, then the sum, S, of an infinite number of terms has an upper limit and is given by the formula:

$$S = \frac{a_1}{1 - r}$$

ILLUSTRATIVE PROBLEMS

1. Find the 15th term of the sequence 50, 46, 42, 38 . . .

Solution $n = 15, d = -4, a_1 = 50$

$$a_n = a_1 + (n-1)d$$
$$a_{15} = 50 + (15-1)(-4)$$
$$a_{15} = 50 + 14(-4)$$
$$a_{15} = 50 - 56$$
$$= -6$$

2. Which term of the series 1, 6, 11 . . . is 96?

Solution $a_1 = 1, d = 5, a_n = 96$

$$a_n = a_1 + (n-1)d$$
$$96 = 1 + 5(n-1)$$
$$96 = 5n - 4$$
$$5n = 100$$
$$n = 20$$

3. Find the sum of the first 10 terms of the series $3 + 5 + 7 \ldots + 21$

Solution $a_1 = 3, n = 10, a_n = 21$

$$S_n = \frac{n}{2}(a_1 + a_n)$$

$$S_{10} = \frac{10}{2}(3 + 21)$$

$$S_{10} = 5(24)$$

$$S_{10} = 120$$

4. Find the sum of the first 20 terms of the series $15, 13\frac{1}{2}, 12 \ldots$

Solution $a_1 = 15, d = -1\frac{1}{2}, n = 20.$

$$S_n = \frac{n}{2}[2a_1 + (n-1)d]$$
$$S_{20} = \frac{20}{2}\left[2 \cdot 15 + (20-1)\left(-1\frac{1}{2}\right)\right]$$
$$S_{20} = 10\left[30 + 19\left(-\frac{3}{2}\right)\right]$$
$$S_{20} = 10\left[30 - \frac{57}{2}\right] = 10\left[\frac{3}{2}\right]$$
$$S_{20} = 15$$

5. Find the sum of all integers between 1 and 100 that are exactly divisible by 9.

Solution   The A.P. is 9, 18, 27 ... 99.

$$a_n = a_1 + (n-1)d$$
$$99 = 9 + 9(n-1)$$
$$n = 11$$

$$S_n = \frac{n}{2}(a_1 + a_n)$$
$$S_{11} = \frac{11}{2}(9 + 99)$$
$$S_{11} = \frac{11}{2}(108)$$
$$S_{11} = 594$$

6. Find the 9th term of the G.P.: 20, 10, 5, $2\frac{1}{2}$ ...

Solution $a_1 + 20, r = \frac{1}{2}, n = 9$

$$a_n = a_1 r^{n-1}$$
$$a_9 = 20\left(\frac{1}{2}\right)^8$$
$$a_9 = \frac{5}{64}$$

7. Find the sum of 5 terms of the G.P.: 27, 9, 3 ...

Solution $r = \frac{1}{3}, a_1 = 27, n = 5$

$$S_n = \frac{a_1 - a_1 r^n}{1 - r}$$
$$S_5 = \frac{27 - 27\left(\frac{1}{3}\right)^5}{1 - \frac{1}{3}}$$
$$S_5 = \frac{27 - \frac{1}{9}}{\frac{2}{3}}$$
$$S_5 = \frac{242}{6} = 40\frac{1}{3}$$

8. Find the sum of the infinite G.P.: 12, 6, 3 ...

Solution $r = \frac{1}{2}, a_1 = 12$

$$S = \frac{a_1}{1-r} = \frac{12}{1-\frac{1}{2}}$$
$$S = \frac{12}{\frac{1}{2}}$$
$$S = 24$$

9. Write the repeating decimal .343434 . . . as a fraction.

Solution▸ Write the number as the sum of an infinite G.P.

$$.343434\ldots = .34 + .0034 + .000034 + \ldots$$

$$a_1 = .34 \text{ and } r = .01$$

$$S = \frac{a_1}{1-r} = \frac{.34}{1-.01}$$

$$S = \frac{.34}{.99}$$

$$S = \frac{34}{99}$$

15. Vectors

Forces and velocities are usually represented as *vectors*. A vector is a quantity having both *magnitude* and *direction*.

We represent a vector by an arrow to show its direction, the length of which is proportional to the magnitude of the vector.

If a vector *a* and a vector *b* react upon an object so that it moves in a new direction, this new vector is called the resultant, or vector sum of *a* and *b*.

In some problems in mechanics we wish to reverse the above procedure; that is, given a vector, we may want to find two perpendicular vectors that, when added, have the given vector as a resultant. These two vectors are called *components* of the given vector.

ILLUSTRATIVE PROBLEMS

1. A plane is flying north at 240 mph when it encounters a west wind blowing east at 70 mph. In what direction will the plane be going and with what speed?

Solution▸

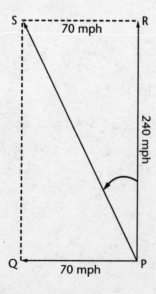

The scale drawing shows vectors for velocities $\overrightarrow{PR}$ and $\overrightarrow{PQ}$. (The arrow is used for a vector.) The vector $\overrightarrow{PS}$ represents the actual path of the plane. It is obtained by completing the parallelogram (or rectangle) *PQSR*. The length of *PS* represents the actual speed of the plane.

$$\overline{PS}^2 = \overline{PR}^2 + \overline{RS}^2$$
$$= 240^2 + 70^2$$
$$= 62,500$$
$$PS = 250 \text{ mph}$$

The bearing angle is

$$\tan \angle RPS = \frac{70}{240} = \frac{7}{24} = 0.2917$$

$$\angle RPS \approx 16°$$

The bearing is N 16° W.

2. A force of 100 lb is acting at 30° to the horizontal. Find the horizontal and vertical components of the given vector.

Solution

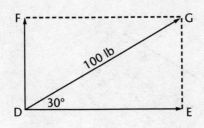

The scale drawing shows the components DE and DF of the given vector DG. From right triangle DEG,

$$\sin 30° = \frac{GE}{100}$$

$$\frac{1}{2} = \frac{GE}{100}$$

$$GE = 50 \text{ lb}$$

$$\cos 30° = \frac{DE}{100}$$

$$\frac{\sqrt{3}}{2} = \frac{DE}{100}$$

$$DE = 50\sqrt{3} \approx 87 \text{ lb}$$

3. In some problems in mechanics, it may be necessary to find the *difference* of two vectors. Given horizontal vector AB and vertical vector AC, find a vector equal to the difference AB − AC. Both vectors have a magnitude of 10.

Solution

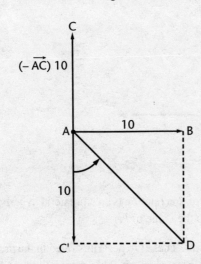

Consider $-\overrightarrow{AC}$ as a vector equal to $\overrightarrow{AC}$ in magnitude and opposite in direction. Thus we are finding the resultant of $\overrightarrow{AB}$ and $-\overrightarrow{AC}$. From right triangle $AC'D$ $\angle CAD = 45°$ and $AD = 10\sqrt{2} \approx 14$. $\overrightarrow{AD}$ has a magnitude of 14 and bearing S 45° E.

16. Variation

Two algebraic functions are applied frequently in science problems. These are generally referred to as *variation* problems.

The variable y is said to vary *directly* as the variable x if $y = kx$ where k represents a constant value. k is usually called the constant of variation or *proportionality constant*.

The graph of this relationship is a straight line passing through the origin. k is equal to the *slope* of the line, where the slope refers to the ratio of the change in y to the change in x.

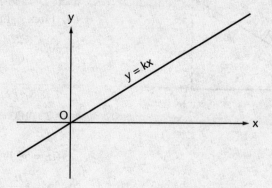

If the volume of an ideal gas is kept constant, the pressure varies *directly* with the temperature; $P = kT$. If one pair of values is given for T and P, k can be determined. In many scientific formulas of this type, the units are frequently defined so that $k = 1$.

The variable y is said to vary *inversely* as x if $y = \frac{k}{x}$ where k is a constant. For example, for several automobiles traveling the same distance, the time t in hours varies inversely as the rate r in miles per hour. In this case the constant of variation is the distance in miles, since $t = \frac{d}{r}$.

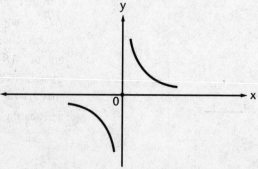

The variation $y = \frac{k}{x}$ may also be written $xy = k$. The graph of this relation is an equilateral hyperbola as shown in the figure. The specific hyperbola will depend on the value of k.

Direct and inverse variations may take several different forms. These are best illustrated in the problems below. If z varies directly as x and y, then $z = kxy$, where k is a constant. This is sometimes called *joint variation*.

If z varies directly as x and inversely as y, then $z = \frac{kx}{y}$, where k is a constant.

16. VARIATION

ILLUSTRATIVE PROBLEMS

1. An auto traveling at a rate of 40 mph covers a distance in 3 hours. At what rate must the auto travel to cover the same distance in 2 hours?

Solution▶

The greater the rate the car travels, the less the time it takes to cover the same distance. Thus, the rate varies *inversely* with time $r = \frac{k}{t}$, where k is the constant of proportionality (distance). Substitute $r = 40$ when $t = 3$. $k = 120$.

$$r = \frac{120}{t} = \frac{120}{2} = 60 \text{ mph}$$

2. If y varies inversely as the square of x, then if x is

(A) multiplied by 2, y is multiplied by 2
(B) increased by 2, y is increased by 4
(C) divided by 2, y is multiplied by 4
(D) decreased by 2, y is increased by 2
(E) multiplied by 2, y is decreased by 2

Solution▶   (C)

$$y = \frac{k}{x^2}$$

Replace x by $\frac{x}{2}$ and let the new value of y be y'.

$$y' = \frac{k}{\left(\frac{x}{2}\right)^2} = \frac{k}{\frac{x^2}{4}} = \frac{4k}{x^2}$$

$$y' = 4\left(\frac{k}{x^2}\right) = 4y$$

When x is divided by 2, y is multiplied by 4.
Likewise, replace x by $2x$.

$$y' = \frac{k}{4x^2} = \frac{1}{4}y. \text{ (A) is not true.}$$

Replace x by $x + 2$.

$$y' = \frac{k}{(x+2)^2} = \frac{k}{x^2 + 4x + 4}. \text{ (B) is not true.}$$

Similarly, (D) and (E) can be shown to be untrue.
The only correct choice is (C).

3. The surface area of a sphere varies directly as the square of a radius. If the area is 36π sq cm when the radius is 3 cm, what is the area when the radius is 5 cm?

Solution▶

$$S = kr^2$$
$$36\pi = 9k$$
$$k = 4\pi$$

Thus, $S = 4\pi r^2$
If $r = 5$,

$$S = 4\pi(5)^2 = 100\pi \text{ sq cm}$$

4. If s varies directly as t^2, what is the constant?

   (A) the product of s and t^2
   (B) the square of s and t
   (C) the quotient of s and t^2
   (D) the sum of s and t^2
   (E) the difference of s and t^2

Solution▶   (C) If $s = kt^2$, then $k = \dfrac{s}{t^2}$

PART FOUR

Math Practice Exercises and Solutions by Topic

CONTENTS

1. Formulas and Linear Equations 81
2. Algebraic Fractions 81
3. Sets 82
4. Functions 83
5. Exponents 83
6. Logarithms 84
7. Equations—Quadratic and Radical 85
8. Inequalities 85
9. Verbal Problems 86
10. Geometry 86
11. Trigonometry 88
12. Graphs and Coordinate Geometry 89
13. Number Systems and Concepts 91
14. Arithmetic and Geometric Progressions 92
15. Vectors 92
16. Variation 93

Solutions to Practice Exercises 94

1. Formulas and Linear Equations

1. If $6x - 18 = 5$, what does $x - 3$ equal?

2. The formula $C = \frac{5}{9}(F - 32)$ converts Fahrenheit readings (F) into Centigrade readings (C). For which temperature are the readings the same?

3. If $V = Bh$ and $B = \pi r^2$, find V in terms of r and h.

4. If $t(z - 3) = k$, what does z equal?

5. Solve for d: $3c - d = 30$ and $5c - 3d = 10$

6. If $7r - 8 = 6 + 7s$, what does $r - s$ equal?

7. If $5p - q = 9$ and $10p - 2q = 7$, then

   (A) $p = q$
   (B) $p > q$
   (C) $p < q$
   (D) $p = q \neq 0$
   (E) cannot be determined from the information given

8. Using the formula $A = \frac{h}{2}(b + c)$, find b in terms of A, h and c.

9. Solve for x and y:

$$3x - 2y = 6$$
$$\frac{y - 1}{x} = \frac{1}{2}$$

(Solutions on page 94)

2. Algebraic Fractions

1. Combine: $\frac{3m}{8} - \frac{m}{4}$

2. Find the capacity of an oil tank if an addition of 15 gal raises the reading from $\frac{1}{4}$ to $\frac{5}{8}$ full.

3. Write the sum of $\frac{2}{x+1}$ and $\frac{3}{x^2-1}$ as a single fraction in simplest form.

4. Find the product: $\frac{x^2-1}{x} \cdot \frac{4x^2}{x+1}$

5. Write the complex fraction $\dfrac{\frac{x}{x+3}}{1 - \frac{x}{x+3}}$ as a simple fraction.

6. Express in simplest form: $\left(\dfrac{3x^2}{x^2-4}\right)\left(\dfrac{x+2}{6x}\right)$

7. Solve for y: $\dfrac{6}{y} + \dfrac{y-3}{2y} = 2$

8. Express as a simple fraction in lowest terms:
$$\dfrac{5y-1}{2y} - \dfrac{5y+2}{3y}$$

9. Express as a single fraction in lowest terms:
$$\dfrac{r^2+r-6}{r^2-1} \div \dfrac{r-2}{r-1}$$

(Solutions on page 95)

3. Sets

1. A is the set of odd numbers between 0 and 6.
   B is the set of whole numbers greater than 1 and less than 6.
   List the members of the set which is the intersection of sets A and B.

2. The solution set of $\sqrt{x+1} - 2 = 0$ is

   (A) $\{1\}$ (B) $\{\}$ (C) $\{3\}$ (D) $\{3,-3\}$ (E) $\{-3\}$

3. Given: set $A = \{a,b,c,d\}$ with a defined operation whose symbol is *. Which statement expresses the fact that b is the identity element for this operation?

   (A) $a*a = b$ (B) $a*b = a$ (C) $a*b = b$ (D) $a*b = c$ (E) $b*c = a$

4. If P and Q are disjoint sets, then

   (A) $P \subset Q$ (B) $P \cap Q = U$ (C) $P \cup Q = \{\}$ (D) $P \cup Q = \{\}$ (E) $Q \subset P$

5. In a school of 1300 students, all students must study either French or Spanish or both. If 800 study French and 700 study Spanish, how many students study both?

6. Let $S = \{a,b,c\}$. How many subsets does it have including itself and the empty set?

7. If $A = \{1,2,3,4,5,6\}$ and $B = \{2,4,6,8,10\}$, how many elements are in $A \cup B$?

(Solutions on page 97)

4. Functions

1. Which relation is a function?

   (A) $\{(x, y) | x^2 + y = 4\}$ (B) $\{(x, y) | x^2 + y^2 = 4\}$
   (C) $\{(x, y) | x^2 - y^2 = 4\}$ (D) $\{(x, y) | x^2 + 4y^2 = 4\}$
   (E) $\{(x, y) | x^2 - 4y^2 = 4\}$

2. If $f(x) = x^2 - 2x + 4$, what is $f(i)$ where $i = \sqrt{-1}$?

3. The function $f(x) = x - x^2$ has its maximum value when x equals

   (A) 1   (B) –1   (C) $\frac{1}{2}$ (D) 0   (E) $-\frac{1}{2}$

4. Express the sum of 3 consecutive even integers as a function of n where n is the smallest integer.

5. If $f(x) = 2x - 3$ and $g(x) = x + 1$, then $f(g(x))$ equals?

6. Write the inverse of the function f as defined by $f(x) = 2x - 3$.

7. Find the largest real range of the function $y = f(x) = 2 + \frac{1}{x}$.

8. The function f is defined as $f(x) = \frac{2x+1}{x-3}$ where $x \neq 3$. Find the value of K so that the inverse of f is $f^{-1}(x) = \frac{3x+1}{x-k}$.

9. If the functions f and g are defined as $f(x) = x^2 - 2$ and $g(x) = 2x + 1$, what is the function $f[g(x)]$?

(Solutions on page 98)

5. Exponents

1. Solve for x: $3^{x+1} - 5 = 22$

2. If the number 0.0000753 is written in the form 7.53×10^n, what is the value of n?

3. Find the solution set of $4^{x-1} = 2^x$.

4. When $x = 27$, what is the value of $(x^{-2})^{1/3}$?

5. If $5^p = 192$, between what two consecutive integers does p lie?

6. The wavelength of violet light is .000016 in. Write this number in scientific notation.

7. Write the numerical value of $r^{2/3} - (4r)^0 + 16r^{-2}$ when $r = 8$.

8. Solve for n: $27^{6-n} = 9^{n-1}$

9. Solve for k: $27^{k+3} = \left(\dfrac{1}{3}\right)^{2-k}$

(Solutions on page 99)

6. Logarithms

1. If $\log x = 1.5877$ and $\log y = 2.8476$, what is the numerical value of $\log x\sqrt[3]{y}$?

2. The expression $\log_b x = 1 + c$ is equivalent to

    (A) $b^{1+c} = x$ (B) $x^{1+c} = b$ (C) $b + bc = x$ (D) $x = (1+c)^b$ (E) $b^{1-x} = c$

3. The expression $\log 2xy$ is equivalent to

    (A) $2(\log x + \log y)$ (B) $2(\log x)(\log y)$ (C) $2\log x + \log y$ (D) $\log 2 + \log x + \log y$
    (E) $\log x + 2\log y$

4. The diagram below represents the graph of which equation?

    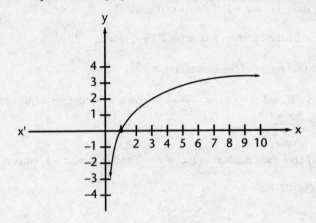

    (A) $y = 2^x$ (B) $y = 10^2$ (C) $y = \log_2 x$ (D) $y = \log_{10} x$ (E) $y = 10\log x$

5. $\log_3 92$ is between what pair of consecutive integers?

6. If $P = K10^{-xt}$, x equals

    (A) $\dfrac{\log k}{t \log P}$ (B) $\dfrac{P10^t}{K}$ (C) $\dfrac{P}{K10^t}$ (D) $\dfrac{\log K - \log P}{t}$ (E) $\dfrac{\log P - \log K}{t}$

7. If $\log_r 6 = S$ and $\log_r 3 = T$, $\log_r\left(\dfrac{r}{2}\right)$ is equal to

    (A) $\dfrac{1}{2}\log_2 r$ (B) $1 - S + T$ (C) $1 - S - T$ (D) $\log_r 2 - 1$ (E) $1 + S + T$

(Solutions on page 100)

7. Equations—Quadratic and Radical

1. For what value of c are the roots of the equation $x^2 + 6x + c = 0$ equal?

2. What is the solution set of $\sqrt{x+1} - 2 = 0$?

3. Find the roots of $2x^2 - 7x + 3 = 0$.

4. If -1 satisfies the equation $x^2 - 3x - k = 0$, what is the value of k?

5. What is the solution set of $x + \sqrt{x-2} = 4$?

6. What is the total number of points whose coordinates satisfy the equations of both $x^2 + y^2 = 4$, and $y = 4$?

7. Solve the system of equations: $3r^2 - rs = 3$ and $6r - s = 10$

8. Solve the equation $s^2 = \dfrac{t}{t+2}$ for t in terms of s.

9. Find, in radical form, the roots of $x^2 - 6x + 7 = 0$.

10. The sum of two numbers is 12 and the sum of their squares is 80. Find the numbers.

(Solutions on page 101)

8. Inequalities

1. If $2x + 2 > 8$, what is the solution set of the inequality?

2. Solve for t: $t - 2 < 3(t - 5)$

3. In the figure, if $90 < q < 180$, what is the range of values x may assume?

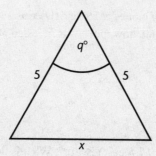

4. $(x + y)^2 < x^2 + y^2$ if

    (A) $x^2 < y^2$ (B) $x < y$ (C) $y < x$ (D) $xy < 0$ (E) $x = y$

5. Find all values of p that satisfy the inequality $-3p + 2 > 11$.

6. If $x > y$ and x, y and z are positive numbers, which of the following is *not* necessarily true?

   (A) $x + z > y + z$ (B) $\dfrac{x}{z} > \dfrac{y}{z}$ (C) $x - z > y - z$ (D) $\dfrac{z}{x} > \dfrac{z}{y}$ (E) $\dfrac{x}{y} > 1$

7. If $5x = 4t$, $6y = 5t$ and $t > 0$, then

   (A) $x > y$ (B) $y > x$ (C) $x = y$ (D) $x = -y$
   (E) Comparison of x and y cannot be determined from the information given.

8. If $t > 1$ and t increases, which of the following decreases?

   (A) $\dfrac{1}{t^2}$ (B) $t - 2$ (C) $t^2 - 4$ (D) $2t - 1$ (E) $2t - 2$

(Solutions on page 103)

9. Verbal Problems

1. In a tank are 50 ℓ water and 20 ℓ of acid. How many ℓ of water must be evaporated to make the solution 40% acid?

2. One pipe takes 6 hours to fill a tank, and another pipe empties it in 8 hours. If both pipes are open, how many hours will it take to fill the tank?

3. Two motorists start from the same point at the same time and travel in opposite directions. One motorist travels 15 mph faster than the other. In 5 hours they are 550 mi apart. Find the rate of speed for each motorist.

4. The sum of the digits of a two-digit number is 10. If 18 is added to the number, the result is equal to the number obtained by reversing the digits of the original number. Find the original number.

5. The length of a rectangle is 2 ft more than its width. If the width were increased by 4 ft and the length diminished by 3 ft, the area would increase by 49 sq ft. Find the dimensions of the rectangle.

6. A woman invested an amount of money at 5% and twice as much at 7%. If her total yearly income from the two investments is $760, how much was invested at each rate?

(Solutions on page 104)

10. Geometry

1. In a circle with center at O, arc ST measures 110°. What is the measure of angle STO?

2. If the angles of a triangle are in the ratio of $2:3:5$, what is the measure of the smallest angle?

3. In the figure $PQ \parallel RS$, find the measure of x.

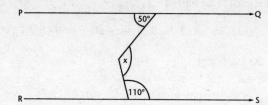

4. The diagonal of a rectangle is 26 cm and its height is 10 cm. Find the area of the rectangle in sq cm.

5. A ship travels 60 mi north, 90 mi west, and then 60 mi north again. How many mi is it from its starting point?

6. Find the length in inches of a tangent drawn to a circle with a 10 in. radius from a point 26 in. from the center of the circle.

7. If the largest possible circular disc is cut from a rectangular piece of tin 8 in. by 12 in., what is the area of the waste tin in square inches in terms of π?

8. A circle is inscribed in a square. What is the ratio of the area of the square to that of the circle in terms of π?

9. At 4:20 P.M., how many degrees has the hour hand of a clock moved since noon?

10. Find the volume of a cube in cu cm if the total surface area of its faces is 150 sq cm.

11. A cylindrical can has a circular base with a diameter of 14 in. and a height of 9 in. Approximately how many gallons does the can hold? (231 cubic in. = 1 gal; use $\pi = \frac{22}{7}$)

12. What is the volume in cubic inches of an open box made by snipping squares 2 in. by 2 in. from the corners of a sheet of metal 8 in. by 11 in. and then folding up the sides?

13. Water 6 in. high in a fish tank 15 in. long by 8 in. wide is poured into a tank 20 in. long by 12 in. wide. What height in inches does it reach in the larger tank?

14. A 6 ft pole is casting a 5 ft shadow at the same time that a flagpole is casting a 22 ft shadow. How many ft high is the flagpole?

15. In the figure, $PQRST$ is a regular pentagon inscribed in the circle. PR and QT are diagonals of the pentagon, forming an angle of $x°$ as shown. What does x equal?

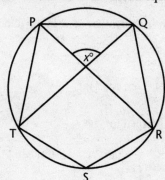

16. A treasure is buried 10 ft from tree T and 12 ft from a straight fence F. If T is 20 ft from F, in how many places may the treasure be buried?

17. If a given statement is true, which of the following statements must also be true?

    (A) the converse of the statement
    (B) the inverse of the statement
    (C) the contrapositive of the statement
    (D) the negative of the statement
    (E) none of these

18. Given point P on a line. In a given plane containing the line, what is the total number of points which are at a distance of 4 units from P and also at a distance of 3 units from the given line?

19. Point Q is 20 cm from plane P in space. What is the locus of points 8 cm from P and 12 cm from point Q?

20. The sum of the measures of the interior angles of a convex polygon is 720°. What is the sum of the measures of the interior angles of a second convex polygon that has two more sides than the first?

(Solutions on page 106)

11. Trigonometry

1. The expression $\dfrac{\sec x}{\tan x \sin x}$ is equivalent to

    (A) $\tan x$ (B) $\dfrac{1}{\cos^2 x}$ (C) $\dfrac{1}{\sin^2 x}$ (D) $\dfrac{\sec^2 x}{\sin x}$ (E) $\cos x$

2. For values of x in the interval $0° \le x \le 360°$, what is the total number of solutions for x in the equation $\cos x(\cos x - 2) = 0$?

3. If $y = \sin x$ and $y = \cos x$ are graphed on the same set of axes and $0 \le x \le 2\pi$, in what quadrants do the graphs intersect?

4. What is the positive value of $\sin(\text{arc tan } \sqrt{2})$?

5. If $\cos \theta = \dfrac{1}{9}$, what is the positive value of $\sin \dfrac{\theta}{2}$?

6. In triangle ABC, $a = 5$, $b = 7$ and $c = 8$. What is the cosine of angle C?

7. Solve the following equation for all values of θ in the interval $0° < \theta < 360°$.

    $$\cos 2\theta + \sin \theta = 0$$

8. Find the value of $\sin(\text{arc sin } 1 + \text{arc cos } 1)$.

9. The expression $\dfrac{2 \tan A}{1 + \tan^2 A}$ is equal to

    (A) $\cos 2A$ (B) $\sin 2A$ (C) $\tan 2A$ (D) $\cot 2A$ (E) $\sec A$

10. If in triangle ABC, m ∠A = 30, $a = \sqrt{5}$, and $b = 4$, angle B

   (A) may be either obtuse or acute
   (B) must be obtuse only
   (C) must be acute only
   (D) may be a right angle
   (E) may be a straight angle

11. Express $\dfrac{\sin 2\theta}{2\sin^2 \theta}$ as a trigonometric function of θ.

12. If $\sin x = \dfrac{3}{5}$ and x is an acute angle, find the value of $\sin 2x$.

13. If x and y are positive acute angles and if $\sin x = \dfrac{4}{5}$ and $\cos y = \dfrac{8}{17}$, what is the value of $\cos(x + y)$?

14. Solve for $\sin x$: $\sqrt{1 - \sin x} = \dfrac{1}{2}$

15. In triangle ABC, $b = 6$, $c = 10$ and m ∠A = 30°. Find the area of the triangle.

16. Express cot 208° as a function of a positive acute angle.

17. For how many values of x in the interval $-\pi \le x \le \pi$ does $2 \sin x = \cos 2x$?

18. What is the numerical value of $\sin \dfrac{\pi}{6} + \cos \dfrac{\pi}{2}$?

19. If m ∠A = 30°, $b = 50$, and $a = 40$, what type of triangle, if any, can be constructed?

   (A) a right triangle only   (B) two distinct triangles
   (C) one obtuse triangle only   (D) no triangle
   (E) an isosceles triangle only

20. The length of a pendulum is 18 cm. Find, in terms of π, the distance through which the tip of the pendulum travels when the pendulum turns through an arc of 120°.

(Solutions on page 109)

12. Graphs and Coordinate Geometry

1. A triangle has as vertices (0, –2), (0, 6) and (5, $4\tfrac{1}{2}$). How many square units are in its area?

2. At what point do the graphs of the equation $y = x - 3$ and $y = 2x - 5$ intersect?

3. A straight line joins the points (0, 1) and (4, 5). For any point (x, y) on this line, write an equation relating y and x.

4. Find the coordinates of the midpoint of the line joining (1, 2) and (5, 8).

5. In the figure, what is the area of the semicircular region in terms of π?

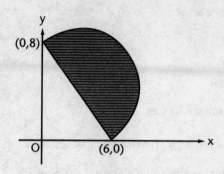

6. In the figure, $PQRS$ is a square of side 6. What are the coordinates of vertex Q?

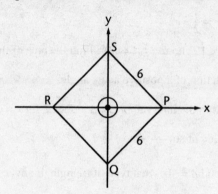

7. In which quadrants does the graph of the equation $xy = 4$ lie?

8. What is the slope of the graph of the equation $3x - 2y = 8$?

9. Write an equation of the axis of symmetry of the graph of the function $f(x) = 2x - x^2$.

10. If the angle of inclination of a straight line is 30°, what is the slope of this line?

11. What is the total number of points whose coordinates satisfy both equations $xy = -6$ and $y = x$?

12. Given the function $\{(x, y) \mid y = x^2 - 2x - 4\}$ whose domain is $-2 \le x \le 4$. If this function is graphed, what are the coordinates of the turning point of the graph?

13. The graph of the relation $x^2 = y^2 + 3$ is

    (A) an ellipse   (B) a parabola   (C) a circle   (D) a hyperbola   (E) a straight line

14. A line L is drawn perpendicular to the graph of the line whose equation is $2x + 3y = 5$. What is the slope of line L?

15. What is the area of the triangle whose vertices have coordinates $(0, 0)$, $(5, 1)$ and $(2, 6)$?

16. What is the y-intercept of the line whose equation is $5x - 2y = 6$?

17. The coordinates of the vertices of $\triangle ABC$ are $A(-2, -2)$, $B(8, -2)$ and $C(1, 6)$. In radical form, find the length of median CM.

18. In problem 17, write an equation of the locus of points equidistant from A and B.

(Solutions on page 113)

13. Number Systems and Concepts

1. The sum of three consecutive *odd* numbers is always divisible by

   (A) 2   (B) 3   (C) 4   (D) 6   (E) 9

2. If s and t are different integers and $\sqrt{st} = 8$, which of the following could *not* be a value of $s + t$?

   (A) 16   (B) 20   (C) 34   (D) 65   (E) none of these

3. If the symbol * is an operation on two numbers, for which of the following definitions of * is the operation commutative?

   (A) $r * s = r - s$ (B) $r * s = \dfrac{r}{s}$ (C) $r * s = \dfrac{r+s}{rs}$ (D) $r * s = \dfrac{r+s}{r}$ (E) $r * s = r^2 - s^2$

4. If $p^2 - 2pq + q^2 = c$ where p and q are positive integers and p is even and q is odd, which of the following is true?

   (A) c is even and a perfect square
   (B) c is odd and a perfect square
   (C) c is even and not a perfect square
   (D) c is odd and not a perfect square
   (E) none of these

5. All of the following numbers may be written as the sum of two prime numbers except one. Which one may *not* be so written?

   (A) 8   (B) 12   (C) 18   (D) 23   (E) 24

6. What is the product of $(1 + i)$ and $(1 + 2i)$?

7. If $i = \sqrt{-1}$, what does $(1 + i)^2$ equal in $a + bi$ form?

8. If $f(x) = x^2 - 3x + 2$, what is $f(i)$ where $i = \sqrt{-1}$?

9. Express in terms of i the sum of $\sqrt{-81}$ and $2\sqrt{-25}$.

10. What complex number is the multiplicative inverse of $2 - i$?

11. Express $\dfrac{2+i}{1-i}$ as a complex number in the form $a + bi$.

12. Find the product of the conjugate complex numbers $(3 + 5i)$ and $(3 - 5i)$.

13. Express $i^5 + i^6$ as a complex number in the form $a + bi$.

14. $\dfrac{3+2i}{3-2i} - \dfrac{3-2i}{3+2i} = ?$

(Solutions on page 116)

14. Arithmetic and Geometric Progressions

1. Find the 20th term of the arithmetic progression 3, 8, 13, 18 . . . .

2. Find the sum of all numbers between 1 and 100 that are divisible by 3.

3. Find the 6th term of the series 1, $-\frac{1}{2}$, $-\frac{1}{4}$ . . . .

4. Find the sum of the first 7 terms of the geometric progression 24, 12, 6 . . . .

5. Find the sum of the infinite geometric progression 3, $\frac{3}{2}$, $\frac{3}{4}$ . . . .

6. Write .151515 . . . as a common fraction.

7. The first row of a theatre has 30 seats, and each succeeding row has 2 additional seats. How many seats are there in the 40th row, which is the last row?

8. In Problem 7, how many seats are there in the entire theatre?

(Solutions on page 118)

15. Vectors

1. Two perpendicular forces of 60 and 80 lb are applied to an object. Find the magnitude of the resultant force and the angle it makes with the larger force.

2. Determine the magnitude and direction of the resultant of two forces, one of 9 newtons whose bearing is S 60° E and the other of 14 newtons whose bearing is S 30° W.

3. A plane is flying south at 120 mph when a west wind starts blowing at 50 mph. Find the magnitude and bearing of the resultant velocity.

4. A steamer leaves New York traveling N 60° E at 30 mph. How fast is it moving northward, and how fast eastward?

5. A man who rows at the rate of 4 mph in still water wishes to row across a river with a current of 3 mph. Find the direction and speed of his boat.

6. In Problem 5, if the man wishes to go straight across the river at 4 mph, in what direction would he have to head and with what speed?

7. An auto weighing 3,000 lb stands on a hill inclined 15° to the horizontal. What force tending to pull it downhill must be overcome by the brakes?

8. A ship is sailing northward in a calm sea at 30 mph. Suddenly a north wind starts blowing at 6 mph, and a current of 10 mph starts moving it eastward. Find the magnitude and direction of the resultant velocity.

(Solutions on page 119)

16. Variation

1. If y varies directly as x and $y = 12$ when $x = 3$, find y when $x = 10$.

2. If y varies inversely as x and $y = 8$ when $x = 3$, find y when $x = 10$.

3. According to Hooke's law, the amount x that a spring is stretched is directly proportional to the force F applied. If a force of $10g$ stretches a spring 3 cm, how large a force will stretch the spring 8 cm?

4. According to Boyle's law, the volume V of an ideal gas at a fixed temperature varies inversely as the pressure P applied to it. If such a gas occupies 20 in.^{3} when under a pressure of 12 lb per in.^{2}, what pressure will result in a volume of 30 in.^{3}?

5. At a fixed voltage, the current I in a DC electric current varies inversely as the resistance R of the circuit. If the resistance is tripled, the current is

    (A) tripled   (B) doubled   (C) divided by 3   (D) divided by 2   (E) divided by 9

6. When an object is dropped from a position above ground, the vertical distance s it falls varies directly as the square of the time t it takes to fall. In 10 seconds, an object falls 1,600 ft. Write a formula expressing s in terms of t.

7. The intensity of illumination on a surface from a source of light varies inversely as the square of the distance of the surface from the source (inverse-square law). The effect of moving a piece of paper three times as far from the source is to

    (A) divide the intensity by 3   (B) multiply the intensity by 3
    (C) divide the intensity by 9   (D) multiply the intensity by 9
    (E) decrease the intensity by 3

8. The frequency of vibration (f) of a pendulum varies inversely as the square root of the length (L) of the pendulum. If a pendulum with a length of 1 ft produces a frequency of 1 vibration per second, write a formula relating f to L.

9. The volume V of an ideal gas varies directly as the absolute temperature T and inversely as the pressure P. If the volume of a gas is 300 cc at 280°K and 560 mm of pressure, write a formula for V in terms of T and P.

10. The gravitational force (F) of attraction between two bodies of masses m and M varies directly as the product of their masses and inversely as the square of the distance d between them. Doubling one of the masses and the distance between the masses will

    (A) double F (B) divide F by 2   (C) multiply F by 4   (D) divide F by 4   (E) multiply F by 2

(Solutions on page 122)

Solutions to Practice Exercises

1. FORMULAS AND LINEAR EQUATIONS

1. $\left(\dfrac{5}{6}\right)$ $6x - 18 = 5$

   Add 18 to both sides.
   $$6x = 23$$
   Divide by 6 on both sides.
   $$x = 3\dfrac{5}{6}$$
   $$3\dfrac{5}{6} - 3 = \dfrac{5}{6}$$

2. $(-40°)$ Let $C = F = x$.
   $$x = \dfrac{5}{9}(x - 32)$$

   Multiply by 9.
   $$9x = 5(x - 32)$$
   $$9x = 5x - 160$$
   Subtract $5x$ from both sides.
   $$4x = -160$$
   $$x = -40°$$

3. $(V = \pi r^2 h)$ $V = Bh,\ B = \pi r^2$

   Substitute πr^2 for B in first equation.
   $$V = \pi r^2 h$$

4. $\left(\dfrac{k}{t} + 3\right)$ $t(z - 3) = k$

   Divide by t.
   $$z - 3 = \dfrac{k}{t}$$
   Add 3 to both sides.
   $$z = \dfrac{k}{t} + 3$$

5. **(30)** Multiply $3c - d = 30$ by -3.
   $$-9c + 3d = -90$$
   $$5c - 3d = 10$$
   Add the two equations.
   $$-4c = -80$$
   $$c = 20$$

Substitute in first equation.

$$3(20) - d = 30$$
$$60 - d = 30$$
$$d = 30$$

6. **(2)**

$$7r - 8 = 6 + 7s$$
$$7r - 7s = 6 + 8$$
$$7(r - s) = 14$$
$$r - s = 2$$

7. **(E)** Multiply $5p - q = 9$ by 2.

$$10p - 2q = 18$$

and

$$10p - 2q = 7$$

The two equations represent *contradictory* statements, and p and q cannot be determined.

8. $\left(\dfrac{2A - ch}{h}\right)$ $\qquad A = \dfrac{h}{2}(b + c)$

Multiply by 2.

$$2A = h(b + c)$$
$$2A = bh + ch$$

Subtract ch.

$$2A - ch = bh$$

Divide by h.

$$b = \dfrac{2A - ch}{h}$$

9. $(x = 4, y = 3)$ $\qquad 3x - 2y = 6$ and $\dfrac{y - 1}{x} = \dfrac{1}{2}$

Cross-multiply $\dfrac{y - 1}{x} = \dfrac{1}{2}$ to find x.

$$2y - 2 = x$$
$$2y = x + 2$$

Substitute in first equation.

$$3x - (x + 2) = 6 \qquad\qquad 2y - 2 = x$$
$$3x - x - 2 = 6 \qquad\qquad 2y - 2 = 4$$
$$2x - 2 = 6 \qquad\qquad 2y = 6$$
$$x = 4 \qquad\qquad y = 3$$
$$x = 4, \quad y = 3$$

2. ALGEBRAIC FRACTIONS

1. $\left(\dfrac{m}{8}\right)$ $\qquad \dfrac{3m}{8} - \dfrac{m}{4}$ L.C.D. $= 8$

$$\dfrac{3m}{8} - \dfrac{2m}{8} = \dfrac{m}{8}$$

2. **(40 gal)** Let x = no. of gal tank holds.

$$\frac{1}{4}x + 15 = \frac{5}{8}x$$

Multiply by 8.

$$2x + 120 = 5x$$
$$120 = 3x$$
$$x = 40$$

3. $\left(\dfrac{2x+1}{x^2-1}\right)$

$$\frac{2}{x+1} + \frac{3}{x^2-1} \quad \text{L.C.D.} = x^2 - 1 = (x+1)(x-1)$$

$$\frac{2(x-1)}{(x+1)(x-1)} + \frac{3}{(x+1)(x-1)} =$$

$$\frac{2x-2+3}{x^2-1} = \frac{2x+1}{x^2-1}$$

4. **(4x(x − 1))**

$$\frac{x^2-1}{x} \cdot \frac{4x^2}{x+1} = \frac{(x+1)(x-1)}{\cancel{x}} \cdot \frac{\cancel{4x^2}^{4x}}{\cancel{x+1}} = 4x(x-1)$$

5. $\left(\dfrac{x}{3}\right)$

$$\frac{\dfrac{x}{x+3}}{1 - \dfrac{x}{x+3}}$$

Multiply numerator and denominator by $x + 3$.

$$\frac{(x+3)\cdot \dfrac{x}{x+3}}{(x+3)\left(1 - \dfrac{x}{x+3}\right)} = \frac{x}{x+3-x} = \frac{x}{3}$$

6. $\left(\dfrac{x}{2(x-2)}\right)$

$$\frac{3x^2}{x^2-4} \cdot \frac{x+2}{6x}$$

$$\frac{\cancel{3x^2}^{x}}{(x+2)(x-2)} \cdot \frac{\cancel{x+2}}{\cancel{6x}_{2}} = \frac{x}{2(x-2)}$$

7. **(3)**

$$\frac{6}{y} + \frac{y-3}{2y} = 2 \quad \text{L.C.D.} = 2y$$

$$(2y)\frac{6}{y} + \frac{(2y)(y-3)}{2y} = 2(2y)$$

$$12 + y - 3 = 4y$$

$$9 = 3y$$

$$y = 3$$

8. $\left(\dfrac{5y-7}{6y}\right)$

$$\frac{5y-1}{2y} - \frac{5y+2}{3y} \quad \text{L.C.D.} = 6y$$

$$\frac{3(5y-1)}{6y} - \frac{2(5y+2)}{6y}$$

$$= \frac{15y - 3 - 10y - 4}{6y} = \frac{5y-7}{6y}$$

9. $\left(\dfrac{r+3}{r+2}\right)$

$$\frac{r^2 + r - 6}{r^2 - 1} \div \frac{r-2}{r-1}$$

$$= \frac{(r+3)(\cancel{r-2})}{(r+1)(\cancel{r-1})} \cdot \frac{\cancel{r-1}}{\cancel{r-2}} = \frac{r+3}{r+1}$$

3. SETS

1. **(3,5)**

   A is the set $\{1,3,5\}$
   B is the set $\{2,3,4,5\}$
   $A \cap B$ is the set $\{3,5\}$

2. **(C)**

$$\sqrt{x+1} - 2 = 0$$
$$\sqrt{x+1} = 2$$
$$x + 1 = 4$$
$$x = 3$$

This value checks in the original equation.

3. **(B)** Since $a * b = a$, b must be the identity element for the operation $*$.

4. **(C)** If P and Q are disjoint sets, they have no elements in common. $P \cap Q = \{\ \}$ (empty set).

5. **(200)**

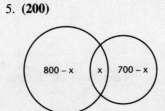

Let x = no. of students studying both.
From the Venn diagram at the left,
$(800 - x) + x + (700 - x) = 1300$
$800 + (700 - x) = 1300$
$1500 - x = 1300$
$x = 200$

6. **(8)** If $S = \{a,b,c\}$, its subsets are $\emptyset$, $\{a\}$, $\{b\}$, $\{c\}$, $\{a,b\}$, $\{a,c\}$, $\{b,c\}$, and S.

7. **(8)** $A \cup B = \{1,2,3,4,5,6,8,10\}$ (8 elements)

4. FUNCTIONS

1. **(A)** For a relation to be a function, only one value of y exists for any given value of x. This is true only for $x^2 + y = 4$, a parabola having a vertical axis of symmetry.

2. **(3 − 2i)** $f(x) = x^2 − 2x + 4$ Substitute i in the equation.

$$f(i) = i^2 − 2i + 4$$
$$= −1 − 2i + 4$$
$$= 3 − 2i$$

3. **(C)** Let $f(c) = x − x^2 = 0$

$$x(1 − x) = 0$$
$$x = 0, x = 1 \text{ (}x\text{-intercepts of parabola)}$$

The axis of symmetry is a vertical line halfway between $x = 0$ and $x = 1$, and the maximum value of $f(x)$ must lie on this line.

4. **(f(n) = 3n + 6)**   Integers are n, $(n + 2)$ and $(n + 4)$.

$$f(n) = n + n + 2 + n + 4$$
$$= 3n + 6$$

5. **(2x − 1)** $f(x) = 2x − 3$ and $g(x) = x + 1$

$$f(g(x)) = 2g(x) − 3$$
$$= 2(x + 1) − 3$$
$$= 2x + 2 − 3$$
$$= 2x − 1$$

6. $\left(\dfrac{x+3}{2}\right)$ $y = f(x) = 2x − 3$

Solve for x.

$$x = \dfrac{y+3}{2}$$

Interchange x and y to get $f^{-1}(x)$.

$$y = \dfrac{x+3}{2}$$

7. **(all real values of y except y = 2)** $y = f(x) = 2 + \dfrac{1}{x}$

The function is defined for all values of x except $x = 0$. Therefore, $\dfrac{1}{x}$ cannot equal 0. The range consists of all real values of y except $y = 2$.

8. **(2)** $y = f(x) = \dfrac{2x+1}{x-3}$ where $x \neq 3$. Interchange x and y.

$$x = \dfrac{2y+1}{y-3}$$

Solve for y.

$$xy - 3x = 2y + 1$$
$$xy - 2y = 3x + 1$$
$$y(x-2) = 3x + 1$$
$$y = \dfrac{3x+1}{x-2}$$

Since $f^{-1}(x) = \dfrac{3x+1}{x-k}$ $K = 2$

9. **($4x^2 + 4x - 1$)** $f(x) = x^2 - 2$ and $g(x) = 2x + 1$

$$f[g(x)] = [g(x)]^2 - 2$$
$$= (2x+1)^2 - 2$$
$$= 4x^2 + 4x + 1 - 2$$
$$= 4x^2 + 4x - 1$$

5. EXPONENTS

1. **(2)**
$$3^{x+1} - 5 = 22$$
$$3^{x+1} = 27$$
$$x + 1 = 3$$
$$x = 2$$

2. **(−5)** $0.0000753 = 7.53 \times 10^n$

   Move the decimal point after the 7, and count back 5 spaces to the original decimal point.

3. **(2)**
$$4^{x-1} = 2^x$$
$$2^{2(x-1)} = 2^x$$
$$2(x-1) = x$$
$$2x - 2 = x$$
$$x = 2$$

4. **$\left(\dfrac{1}{9}\right)$**
$$\left(x^{-2}\right)^{1/3} = x^{-2/3}$$
$$= \dfrac{1}{x^{2/3}} = \dfrac{1}{27^{2/3}}$$
$$= \dfrac{1}{(27^{1/3})^2} = \dfrac{1}{3^2} = \dfrac{1}{9}$$

5. **(3 and 4)** $5^p = 192$; $5^3 = 125$ and $5^4 = 625$, p is between 3 and 4

6. **(1.6×10^{-5})** Write .000016 as 1.6×10^{-n}. Since the decimal point is moved 5 places, the number is 1.6×10^{-5}

7. $\left(3\dfrac{1}{4}\right)$

$$r^{2/3} - (4r)^0 + 16r^{-2}$$
$$= 8^{2/3} - (4 \cdot 8)^0 + \left(16 \cdot 8^{-2}\right)$$
$$= 4 - 1 + \left(16 \cdot \dfrac{1}{64}\right)$$
$$= 3 + \dfrac{1}{4} = 3\dfrac{1}{4}$$

8. **(4)**
$$27^{6-n} = 9^{n-1}$$
$$3^{3(6-n)} = 3^{2(n-1)}$$
$$3(6-n) = 2(n-1)$$
$$18 - 3n = 2n - 2$$
$$20 = 5n$$
$$n = 4$$

9. $\left(-5\dfrac{1}{2}\right)$
$$27^{k+3} = \left(\dfrac{1}{3}\right)^{2-k}$$
$$3^{3(k+3)} = 3^{-(2-k)}$$
$$3(k+3) = -(2-k)$$
$$3k + 9 = -2 + k$$
$$2k = -11$$
$$k = -5\dfrac{1}{2}$$

6. LOGARITHMS

1. **(2.5369)**
$$\log x^3\sqrt{y} = \log x + \dfrac{1}{3}\log y$$
$$= 1.5877 + \dfrac{1}{3}(2.8476)$$
$$= 1.5877 + .9492$$
$$= 2.5369$$

2. **(A)**
$$\log_b x = 1 + c$$
$$b^{1+c} = x$$

3. **(D)** $\log 2xy = \log 2 + \log x + \log y$

4. **(C)** From the graph, $2^0 = 1, 2^1 = 2, 2^2 = 4$ and $2^3 = 8$ where y is the exponent and x is the power of 2. Thus, $2^y = x$ or $y = \log_2 x$

5. **(4 and 5)** Let $\log_3 92 = x$, then $3^x = 92$.

Since $3^4 = 81$ and $3^5 = 243$, x is between 4 and 5.

6. **(D)** $P = K10^{-xt}$

$$\log P = \log K = xt \log 10$$
$$\log P = \log K - xt$$
$$xt = \log K - \log P$$
$$x = \frac{\log K - \log P}{t}$$

7. **(B)** If $\log_r 6 = S$, $\log_r 3 = T$, then $\log_r \frac{6}{3} = \log_r 2 = S - T$.

$$\log_r\left(\frac{r}{2}\right) = \log_r r - \log_r 2$$
$$= 1 - (S - T)$$
$$= 1 - S + T$$

7. EQUATIONS—QUADRATIC AND RADICAL

1. **(9)** $x^2 + 6x + c = 0$

$$b^2 - 4ac = 36 - 4c = 0$$
$$4c = 36$$
$$c = 9$$

2. **(3)**

$$\sqrt{x+1} - 2 = 0$$
$$\sqrt{x+1} = 2$$
$$x + 1 = 4$$
$$x = 3$$

3. $\left(3, \frac{1}{2}\right)$

$$2x^2 - 7x + 3 = 0$$
$$(2x - 1)(x - 3) = 0$$

$2x - 1 = 0$ | $x - 3 = 0$
$x = \frac{1}{2}$ | $x = 3$

4. **(4)**

$$x^2 - 3x - k = 0$$
$$(-1)^2 - 3(-1) - k = 0$$
$$1 + 3 - k = 0$$
$$k = 4$$

5. **(3)**
$$x + \sqrt{x-2} = 4$$
$$\sqrt{x-2} = 4 - x$$
$$x - 2 = 16 - 8x + x^2$$
$$x^2 - 9x + 18 = 0$$
$$(x-3)(x-6) = 0$$

$x - 3 = 0$	$x - 6 = 0$
$x = 3$	$x = 6$

Check both values in original equation.

$x + \sqrt{x-2} = 4$	$x + \sqrt{x-2} = 4$
$3 + \sqrt{3-2} = 4$	$6 + \sqrt{6-2} = 4$
$3 + 1 = 4$ (checks)	$6 + 2 = 4$ (does not check)

$x = 3$ is the only root

6. **(0)** $x^2 + y^2 = 4$ graphs as a circle of radius 2 with center at the origin.
$y = 4$ graphs as a horizontal line 4 units above the x-axis.
There are no intersections.

7.

r	3	$\frac{1}{3}$
s	8	−8

$3r^2 - rs = 3$

$6r - s = 10$ or $s = 6r - 10$
Substitute this value of s in original equation.

$$3r^2 - r(6r - 10) = 3$$
$$3r^2 - 6r^2 + 10r - 3 = 0$$
$$-3r^2 + 10r - 3 = 0$$
$$3r^2 - 10r + 3 = 0$$
$$(3r - 1)(r - 3) = 0$$

$r = \frac{1}{3}$	$r = 3$
$3r - 1 = 0$	$r - 3 = 0$
$s = 6r - 10$	$s = 6r - 10$
$= 6\left(\frac{1}{3}\right) - 10$	$s = 6(3) - 10$
$= -8$	$s = 8$

$r = \frac{1}{3}, s = -8$, and $r = 3, s = 8$

8. $\left(t = \dfrac{2s^2}{1-s^2}\right)$

$$s^2 = \dfrac{t}{t+2}$$
$$s^2(t+2) = t$$
$$s^2 t + 2s^2 = t$$
$$2s^2 = t - s^2 t$$
$$2s^2 = t(1-s^2)$$
$$t = \dfrac{2s^2}{1-s^2}$$

9. $(3 \pm \sqrt{2})$ $x^2 - 6x + 7 = 0$

$$x = \dfrac{-b \pm \sqrt{b^2 - 4ac}}{2a}$$

Substitute $a = 1$, $b = -6$, $c = 7$.

$$x = \dfrac{6 \pm \sqrt{36 - 4(1 \cdot 7)}}{2} = \dfrac{6 \pm \sqrt{36 - 28}}{2}$$
$$= \dfrac{6 \pm \sqrt{8}}{2} = \dfrac{6 \pm 2\sqrt{2}}{2}$$
$$= 3 \pm \sqrt{2}$$

10. **(8,4)** $x + y = 12$ or $y = 12 - x$
$x^2 + y^2 = 80$
Substitute $y = 12 - x$.

$$x^2 + (12 - x)^2 = 80$$
$$x^2 + 144 - 24x + x^2 = 80$$
$$2x^2 - 24x + 64 = 0$$
$$x^2 - 12x + 32 = 0$$
$$(x - 8)(x - 4) = 0$$

$x - 8 = 0$	$x - 4 = 0$
$x = 8$	$x = 4$

8. INEQUALITIES

1. $(x > 3)$

$$2x + 2 > 8$$
$$2x > 6$$
$$x > 3$$

2. $\left(t > 6\dfrac{1}{2}\right)$

$$t - 2 < 3(t - 5)$$
$$t - 2 < 3t - 15$$
$$t - 3t < 2 - 15$$
$$-2t < -13$$
$$t > 6\dfrac{1}{2}$$

3. $(5\sqrt{2} < x < 10)$

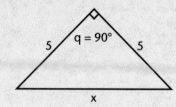

When $q = 90°$, $x = 5\sqrt{2}$
When $q = 180°$, $x = 10$
When $90° < q < 180°$, $5\sqrt{2} < x < 10$

4. **(D)**

$$(x+y)^2 < x^2 + y^2$$
$$x^2 + 2xy + y^2 < x^2 + y^2$$
$$2xy < 0$$
$$xy < 0$$

5. $(p < -3)$

$$-3p + 2 > 11$$
$$-3p > 9$$
$$p < -3$$

6. **(D)** If $x > y$, in (A), (B), and (C), we are adding, dividing, and subtracting z from both sides of the inequality; none of these operations affect the inequality when z is a positive number. In (E), since $x > y$ and they are both positive numbers, $\frac{x}{y}$ must be greater than 1. In (D), we are inverting the two fractions in (B), and this changes the inequality.

7. **(B)**

$$5x = 4t, \quad 6y = 5t, \quad t > 0$$

$$x = \frac{4}{5}t, \quad y = \frac{5}{6}t$$

Since $t > 0$, $y > x$

8. **(A)** (B), (C), (D), and (E) all show increases as t increases when $t > 1$. In (A) t is in the denominator, and $\frac{1}{t^2}$ decreases as t increases beyond 1.

9. VERBAL PROBLEMS

1. **(20)** Let x = no. of ℓ of water to be evaporated.

$$.40(70 - x) = 20$$
$$28 - .4x = 20$$
$$.4x = 8$$
$$x = 20 \, \ell$$

2. **(24)** Let x = no. of hours to fill the tank.

$$\frac{x}{6} - \frac{x}{8} = 1 \quad \text{L.C.D.} = 24$$
$$4x - 3x = 24$$
$$x = 24 \text{ hours}$$

3. $\left(47\frac{1}{2} \text{ mph and } 62\frac{1}{2} \text{ mph}\right)$

$$\begin{array}{c|c}
r \text{ mph} & (r+15) \text{ mph} \\
\hline
5r \text{ mi} \quad 0 & 5(r+15) \text{ mi}
\end{array}$$

$$5r + 5(r+15) = 550$$
$$5r + 5r + 75 = 550$$
$$10r = 475$$
$$r = 47\frac{1}{2} \text{ mph}$$
$$r + 15 = 62\frac{1}{2} \text{ mph}$$

4. **(46)** Let t = ten's digit of original number
u = unit's digit of original number
$10t + u$ = original number
$10u + t$ = number with reversed digits
$t + u = 10$

$$10t + u + 18 = 10u + t$$
$$9t - 9u = -18$$
$$t - u = -2$$
$$\underline{t + u = 10} \text{ (from the given information)}$$
$$2t = 8$$
$$t = 4$$
$$u = 6$$

The original number is 46.

5. **(53, 55)**

[Diagram: rectangle with width x and length $x+2$, area $x(x+2)$; second rectangle with width $x-1$ and length $x+4$, area $(x-1)(x+4)$]

$$(x+4)(x-1) = x(x+2) + 49$$
$$x^2 + 3x - 4 = x^2 + 2x + 49$$
$$x = 53$$
$$x + 2 = 55$$

6. **($4000 at 5%, $8000 at 7%)**

Let x = amount invested at 5%
$y = 2x$ = amount invested at 7%
$$.05x + .07(2x) = 760$$
$$.05x + .14x = 760$$
$$.19x = 760$$
$$19x = 76000$$
$$x = \$4000$$
$$y = \$8000$$

10. GEOMETRY

1. **(35°)**

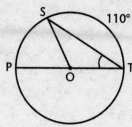

   Arc $SOT = 110°$
   Arc $PS = 180° - 110° = 70°$

   $\angle STO = \dfrac{1}{2}$ arc $PS = \dfrac{1}{2}(70°) = 35°$

2. **(36°)** Let $2x$ = smallest angle in degrees, $3x$ and $5x$ = other two angles in degrees.

$$2x + 3x + 5x = 180°$$
$$10x = 180°$$
$$2x = 36°$$

3. **(120°)**

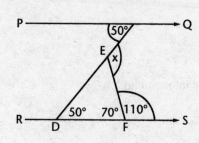

   $\angle EDF = 50°$ and $\angle EFD = 70°$ (alternate interior angles).
   Angle x is an exterior angle of $\triangle DEF$, and is equal to the sum of the two remote interior angles.

$$x = 50° + 70° = 120°$$

4. **(240)**

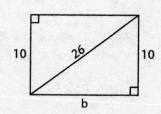

   $b^2 + 10^2 = 26^2$
   $b^2 + 100 = 676$
   $b^2 = 576$
   $b = 24$
   Area $= bh$
   $= 24 \times 10 = 240$ sq cm

5. **(150)**

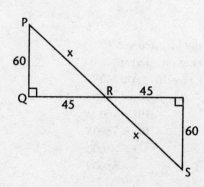

   In $\triangle PQR$, $PQ = 60 = 15(4)$
   $QR = 45 = 15(3)$
   $PR = 15(5) = 75$ (3–4–5 right $\triangle$)
   $PS = 2x = 150$ mi

6. **(24)** 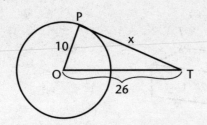 The tangent is perpendicular to the radius at the point of contact. In right triangle *OTP*

$$x^2 + 10^2 = 26^2$$
$$x^2 + 100 = 676$$
$$x^2 = 576$$
$$x = 24 \text{ in.}$$

7. **(96 − 16π)**

Area of rectangle = 12 × 8 = 96 sq in.
Area of circle = $\pi \cdot 4^2$ = 16π sq in.
Area of waste tin = (96 − 16π) sq in.

8. $\left(\dfrac{4}{\pi}\right)$

Let radius of circle be 1.
Side of square is 1 + 1 = 2.
Area of square = 2^2 = 4
Area of circle = $\pi \cdot 1^2 = \pi$
Ratio = $\dfrac{4}{\pi}$

9. **(130°)** From noon to 4:00 P.M., the hour hand has moved from 12 to 4: $\frac{1}{3}$ of 360° = 120°.

In the next 20 minutes it moves $\frac{1}{3}$ of the distance from 4:00 to 5:00: $\frac{1}{3}$ of 30° = 10°.

120° + 10° = 130°

10. **(125)**

Surface area = $6s^2$ = 150 sq cm
$s^2 = 25$
$s = 5$
Volume = $s^3 = 5^3$ = 125 cu cm

11. **(6)**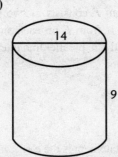

$r = 7, h = 9, V = \pi r^2 h$

$V = \dfrac{22}{7}(49)(9) = 1386$ cu in.

$\dfrac{1386}{231} = 6$ gal

12. **(56)**

Volume = 7 × 4 × 2
     = 56 cu in.

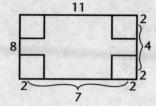   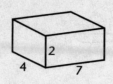

13. **(3)**

$$15 \times 8 \times 6 = 12 \times 20 \times h$$
$$36 = 12h$$
$$h = 3 \text{ in.}$$

14. $\left(26\dfrac{2}{5}\right)$

$$\frac{x}{22} = \frac{6}{5}$$

Cross-multiply.

$$5x = 132$$
$$x = 26\frac{2}{5} \text{ ft}$$

15. **(108°)** $\widehat{PQ} = \widehat{TS} = \widehat{RS} = \dfrac{1}{5}(360°) = 72°$

$$x = \frac{1}{2}(\widehat{PQ} + \widehat{TR})$$
$$x = \frac{1}{2}(72° + 144°)$$
$$= 36° + 72° = 108°$$

16. **(2)**

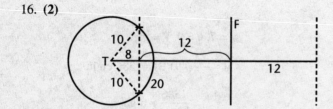

The locus of points 10 ft. from T is a circle of radius 10. The locus of points 12 ft from F consists of two parallel lines 12 ft from F on each side. The circle and one parallel line intersect in two points.

17. **(C)** The only statement that has the same truth value as the given statement is the *contrapositive* of the original statement. This is the *converse* of the *inverse* of the original statement.

18. **(4)** 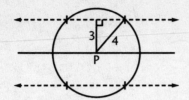 The locus of points 4 units from P is a circle of radius 4 with center at P. The locus of points 3 units from the given line consists of two parallel lines 3 units from the line. The two parallel lines intersect the circle in 4 points.

19. **(1)** 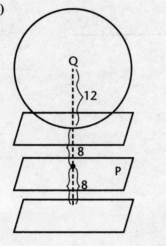 The locus of points 8 cm from P consists of two planes parallel to P and 8 cm from it. The locus of points 12 cm from Q is a sphere of radius 12 and center at Q. The sphere intersects one of the parallel planes in one point (point of tangency).

20. **(1080°)** The sum of the interior angles of an n-sided polygon is $(n-2)180° = 720°$. Divide by 180.

$$n - 2 = 4$$
$$n = 6$$

The second polygon has $6 + 2 = 8$ sides.
Sum of its interior angles $= 180(8 - 2)$
$\qquad\qquad\qquad\qquad\quad = 180(6) = 1080°$

II. TRIGONOMETRY

1. **(C)**

$$\frac{\sec x}{\tan x \sin x} = \frac{\frac{1}{\cos x}}{\frac{\sin x \sin x}{\cos x}}$$

$$= \frac{1}{\sin^2 x}$$

2. **(2)** $\cos x(\cos x - 2) = 0$

$\cos x = 0$	$\cos x - 2 = 0$
$x = \dfrac{\pi}{2}, \dfrac{3\pi}{2}$	$\cos x = 2$
2 solutions	no solutions

3. **(I and III)** Curves intersect in Quadrants I and III.

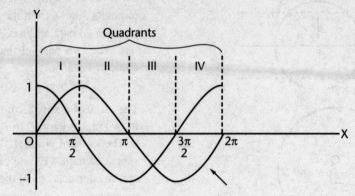

4. $\left(\sqrt{\dfrac{2}{3}}\right)$ $\sin(\arctan \sqrt{2})$. Let $x = \arctan \sqrt{2}$.

$$\tan x = \sqrt{2}$$
$$\sin x = \dfrac{\sqrt{2}}{\sqrt{3}} = \sqrt{\dfrac{2}{3}}$$

5. $\left(\dfrac{2}{3}\right)$ $\sin \dfrac{\theta}{2} = \sqrt{\dfrac{1-\cos\theta}{2}}$

$$= \sqrt{\dfrac{1-\tfrac{1}{9}}{2}} = \sqrt{\dfrac{\tfrac{8}{9}}{2}}$$
$$= \sqrt{\dfrac{4}{9}} = \dfrac{2}{3}$$

6. $\left(\dfrac{1}{7}\right)$ $c^2 = a^2 + b^2 - 2ab\cos C$

$$64 = 25 + 49 - 2(35)\cos C$$
$$70\cos C = 25 + 49 - 64 = 10$$
$$\cos C = \dfrac{10}{70} = \dfrac{1}{7}$$

7. **(90°, 210°, 330°)**

$$\cos 2\theta + \sin\theta = 0$$
$$1 - 2\sin^2\theta + \sin\theta = 0$$
$$2\sin^2\theta - \sin\theta - 1 = 0$$
$$(2\sin\theta + 1)(\sin\theta - 1) = 0$$

$2\sin\theta + 1 = 0$ | $\sin\theta = 1$
$\sin\theta = -\dfrac{1}{2}$ | $\theta = 90°$
$\theta = 210°, 330°$

3 values: 90°, 210°, 330°

8. **(1)** Let $x = \text{arc sin } 1$, $y = \text{arc cos } 1$.
$\sin x = 1, x = 90°$
$\cos y = 1, y = 0°$

$$\sin(\text{arc sin } 1 + \text{arc cos } 1)$$
$$= \sin(x + y)$$
$$= \sin x \cos y + \cos x \sin y$$
$$= \sin 90° \cos 0° + \cos 90° \sin 0°$$
$$= 1 \cdot 1 + 0 \cdot 0 = 1$$

9. **(B)**
$$\frac{2 \tan A}{1 + \tan^2 A} = \frac{\frac{2 \sin A}{\cos A}}{1 + \frac{\sin^2 A}{\cos^2 A}}$$

$$= \frac{\frac{2 \sin A}{\cos A}}{\frac{\cos^2 A + \sin^2 A}{\cos^2 A}} = \frac{\frac{2 \sin A}{\cos A}}{\frac{1}{\cos^2 A}}$$

$$= \frac{2 \sin A}{\cos A} \cdot \frac{\cos^2 A}{1} = 2 \sin A \cos A = \sin 2A$$

10. **(A)** Use the law of sines to find angle B.

$$\frac{a}{\sin A} = \frac{b}{\sin B}$$

$$\frac{\sqrt{5}}{\sin 30°} = \frac{4}{\sin B}$$

$$\sqrt{5} \sin B = 4 \sin 30° = 4\left(\frac{1}{2}\right) = 2$$

$$\sin B = \frac{2}{\sqrt{5}} \cdot \frac{\sqrt{5}}{\sqrt{5}} = \frac{2\sqrt{5}}{5} < 1$$

Thus B may be acute or obtuse, and since $a < b$, $\triangle ABC$ may be either obtuse or acute.

11. **(cot θ)**
$$\frac{\sin 2\theta}{2\sin^2 \theta} = \frac{2 \sin \theta \cos \theta}{2 \sin \theta} = \cot \theta$$

12. $\left(\dfrac{24}{25}\right)$

$\sin x = \dfrac{3}{5}$, $\cos x = \dfrac{4}{5}$

$\sin 2x = 2 \sin x \cos x$

$= 2 \cdot \dfrac{3}{5} \cdot \dfrac{4}{5} = \dfrac{24}{25}$

13. $\left(-\dfrac{36}{85}\right)$

$\cos(x+y) = \cos x \cos y - \sin x \sin y$

$= \dfrac{3}{5} \cdot \dfrac{8}{17} - \dfrac{4}{5} \cdot \dfrac{15}{17}$

$= \dfrac{24}{85} - \dfrac{60}{85} = -\dfrac{36}{85}$

14. $\left(\dfrac{3}{4}\right)$

$\sqrt{1 - \sin x} = \dfrac{1}{2}$

Square both sides.

$1 - \sin x = \dfrac{1}{4}$

$-\sin x = -\dfrac{3}{4}$

$\sin x = \dfrac{3}{4}$

Substitute in original equation.

$\sqrt{1 - \sin x} = \sqrt{1 - \dfrac{3}{4}} = \sqrt{\dfrac{1}{4}} = \dfrac{1}{2}$

15. **(15)**

$\text{Area} = \dfrac{1}{2} bc \sin A$

$= \dfrac{1}{2} \cdot (6) \cdot (10) \cdot \sin 30°$

$= \dfrac{1}{2} \cdot (60) \cdot \dfrac{1}{2} = 15$

16. **(cot 28°)** cot x is positive in Quadrant III, so that cot 208° is positive. cot 208° = cot(208° − 180°) = cot 28°

17. **(2)**

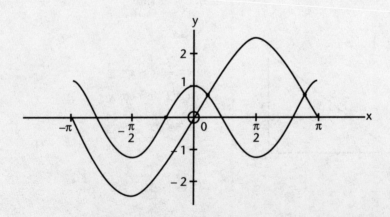

Graph $y = 2 \sin x$ and $y = \cos 2x$ on same set of axes in interval $-\pi \le x \le \pi$. Graph shows 2 values.

18. $\left(\dfrac{1}{2}\right)$ $\sin\dfrac{\pi}{6} + \cos\dfrac{\pi}{2} = \dfrac{1}{2} + 0 = \dfrac{1}{2}$

19. **(B)** Use law of sines.

$$\dfrac{a}{\sin A} = \dfrac{b}{\sin B}$$

$$\dfrac{40}{\sin 30} = \dfrac{50}{\sin B}$$

$$\dfrac{40}{\dfrac{1}{2}} = \dfrac{50}{\sin B}$$

$$40 \sin B = 25$$

$$\sin B = \dfrac{25}{40} = \dfrac{5}{8}$$

B may have two values, one acute or one obtuse. Also, since $a < b$, two triangles may be constructed.

20. **(12π)** $120° = 120 \times \dfrac{\pi}{180} = \dfrac{2\pi}{3}$ radians

$$s = r\theta \ (\theta \text{ in radians})$$

$$s = 18\left(\dfrac{2\pi}{3}\right) = 12\pi \text{ cm}$$

12. GRAPHS AND COORDINATE GEOMETRY

1. **(20)**

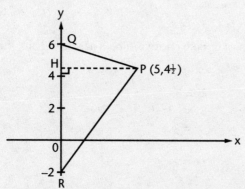

In $\triangle PQR$, use $QR = 8$ as base and $PH = 5$ as altitude.

$$\text{Area} = \dfrac{1}{2}bh = \dfrac{1}{2}(8)(5) = 20$$

2. **(2, –1)** Solve the two equations simultaneously.

$$y = x - 3, \qquad y = 2x - 5$$
$$2x - 5 = x - 3$$
$$x = 2$$
$$y = 2 - 3 = -1$$

3. $(y = x + 1)$ Slope of line:
$$m = \frac{5-1}{4-0} = \frac{4}{4} = 1$$
$$\frac{y-1}{x-0} = 1$$
$$y - 1 = x$$
$$y = x + 1$$

4. **(3,5)** Let (x, y) be the coordinates of the midpoint.
$$x = \frac{1}{2}(1+5) = 3 \qquad y = \frac{1}{2}(2+8) = 5$$
Coordinates are (3,5).

5. $\left(\frac{25\pi}{2}\right)$ The diameter of the semi-circle is the hypotenuse of a 6–8–10 triangle. Radius of the semi-circle is 5.
$$\text{Area} = \frac{1}{2}\pi r^2 = \frac{1}{2}\pi(5)^2 = \frac{25\pi}{2}$$

6. $(0, -3\sqrt{2})$ Let $OQ = OP = x$.
$$x^2 + x^2 = 6^2$$
$$2x^2 = 36$$
$$x^2 = 18$$
$$x = \sqrt{18} = 3\sqrt{2}$$
Coordinates of Q are $(0, -3\sqrt{2})$.

7. **(QI, QIII)** Since xy is positive, x and y must be both positive or both negative. Thus, the graph is in quadrants I and III.

8. $\left(\frac{3}{2}\right)$
$$3x - 2y = 8$$
$$-2y = -3x + 8$$
$$y = \frac{3}{2}x - 4$$
$$\text{slope} = \frac{3}{2}$$

9. **($x = 1$)** $y = f(x) = 2x - x^2$

$$\text{Let } y = 2x - x^2 = 0$$
$$x(2 - x) = 0$$
$$x = 0, \quad x = 2 \text{ (x-intercepts)}$$

The axis of symmetry is a vertical line halfway between $x = 0$ and $x = 2$. Its equation is $x = 1$.

10. $\left(\dfrac{1}{\sqrt{3}}\right)$ Slope = tangent of angle of inclination

$$= \tan 30° = \frac{1}{\sqrt{3}}$$

11. **(0)** The line $y = x$ lies in QI and QIII.
    The curve $xy = -6$ lies in QII and QIV.
    There are *no* intersections.

12. **(1, –5)** Find the x-intercepts of the parabola.
    Let $x^2 - 2x - 4 = 0$.

$$x = \frac{2 \pm \sqrt{4 + 16}}{2} = \frac{2 \pm \sqrt{20}}{2}$$
$$= \frac{2 \pm 2\sqrt{5}}{2} = 1 \pm \sqrt{5}$$

The axis of symmetry is a vertical line midway between the two intercepts. Its equation is $x = 1$.
Substitute in the original equation $y = x^2 - 2x - 4$.
$$y = 1 - 2 - 4 = -5.$$
Turning point is (1, –5).

13. **(D)** $x^2 = y^2 + 3$ or $x^2 - y^2 = 3$, which is a hyperbola.

14. $\left(\dfrac{3}{2}\right)$ Find slope of $2x + 3y = 5$. Solve for y.

$$y = -\frac{2}{3}x + \frac{5}{3}. \quad\quad\quad \text{Slope is } -\frac{2}{3}.$$

Slope of perpendicular line is negative reciprocal of $-\dfrac{2}{3}$, or $\dfrac{3}{2}$.

15. **(14)**

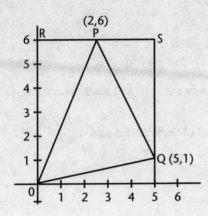

Box $\triangle OPQ$ into rectangle $ORST$ as shown.

$\triangle OPQ = \text{Rect } ORST - (\triangle ORP + \triangle PSQ + \triangle OQT)$

$\text{Rect } ORST = 6 \times 5 = 30$

$\triangle ORP = \frac{1}{2}(6 \times 2) = 6$

$\triangle PSQ = \frac{1}{2}(5 \times 3) = 7\frac{1}{2}$

$\triangle OQT = \frac{1}{2}(5 \times 1) = 2\frac{1}{2}$

$\phantom{\triangle OQT = \frac{1}{2}(5 \times 1) =}\overline{\text{Total} = 16}$

$\triangle OPQ = 30 - 16 = 14$

16. **(−3)**

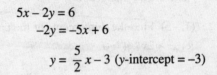

$5x - 2y = 6$

$-2y = -5x + 6$

$y = \frac{5}{2}x - 3$ (y-intercept = −3)

17. **$(2\sqrt{17})$**

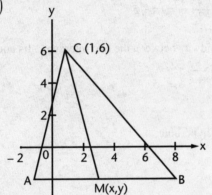

$x = \frac{1}{2}(-2 + 8) = 3$

$y = -2$

Coordinates of M are $(3, -2)$

$CM = \sqrt{(3-1)^2 + (-2-6)^2}$

$ = \sqrt{4 + 64}$

$ = \sqrt{68} = 2\sqrt{17}$

18. **($x = 3$)** Locus is vertical line through M whose equation is $x = 3$.

13. NUMBER SYSTEMS AND CONCEPTS

1. **(B)** Let the numbers be $2n + 1$, $2n + 3$, and $2n + 5$.

$$\text{Sum} = 6n + 9$$
$$= 3(2n + 3)$$

Sum is always divisible by 3.

2. **(A)** If $\sqrt{st} = 8$, $st = 64$, $s \neq t$ (integers)

Therefore, s and t cannot both be 8, or $s + t$ cannot be 16.

3. **(C)** Subtraction and division are not commutative operations, eliminating (A), (B), and (E).

   Using (D) for *, $r*s = \dfrac{r+s}{r}$ and $s*r = \dfrac{s+r}{s}$

   Since these two fractions are *not* equal, (D) is not commutative. Now try (C).

   $$r*s = \dfrac{r+s}{rs} \text{ and } s*r = \dfrac{s+r}{sr}$$

   These two fractions are equal and (C) is a commutative operation.

4. **(B)** $p^2 - 2pq + q^2 = (p-q)^2 = c$

   c is a perfect square. Since p is even and q is odd, $(p-q)$ is odd and so is $(p-q)^2$.

5. **(D)** $8 = 5 + 3$; $12 = 7 + 5$; $18 = 11 + 7$; $24 = 19 + 5$; but 23 cannot be written as the sum of two prime numbers.

6. **(−1 + 3i)**

   $$(1+i)(1+2i) = 1 + 2i + i + 2i^2$$
   $$= 1 + 3i + 2(-1)$$
   $$= -1 + 3i$$

7. **(2i)**

   $$(1+i)^2 = 1 + 2i + i^2$$
   $$= 1 + 2i - 1 = 2i$$

8. **(1 − 3i)**

   $$f(x) = x^2 - 3x + 2$$
   $$f(i) = i^2 - 3i + 2$$
   $$= -1 - 3i + 2$$
   $$= 1 - 3i$$

9. **(19i)**

   $$\sqrt{-81} + 2\sqrt{-25} = 9i + 2(5i)$$
   $$= 9i + 10i$$
   $$= 19i$$

10. $\left(\dfrac{2}{5} + \dfrac{1}{5}i\right)$ Let $a + bi$ = multiplicative inverse of $(2 - i)$.

    $$a + bi = \dfrac{1}{2-i} \cdot \dfrac{2+i}{2+i} = \dfrac{2+i}{4-i^2}$$
    $$= \dfrac{2+i}{4-(-1)} = \dfrac{2+i}{5}$$
    $$= \dfrac{2}{5} + \dfrac{1}{5}i$$

11. $\left(\dfrac{1}{2} + \dfrac{3}{2}i\right)$

    $$\dfrac{2+i}{1-i} \cdot \dfrac{1+i}{1+i} = \dfrac{2+3i+i^2}{1-i^2}$$
    $$= \dfrac{2+3i-1}{1-i^2} = \dfrac{1+3i}{2}$$
    $$= \dfrac{1}{2} + \dfrac{3}{2}i$$

12. **(34)**
$$(3+5i)(3-5i) = 9 - 25i^2$$
$$= 9 - 25(-1)$$
$$= 9 + 25 = 34$$

13. **(−1 + i)** $i^2 = -1$; $i^4 = 1$; $i^5 = i^4 \cdot i = i$; $i^6 = i^2 \cdot i^4 = (-1)(1) = -1$
$i^5 + i^6 = i - 1$ or $-1 + i$

14. $\left(\dfrac{24i}{13}\right)$
$$\frac{3+2i}{3-2i} - \frac{3-2i}{3+2i} = \frac{(3+2i)^2 - (3-2i)^2}{9 - 4i^2}$$
$$= \frac{(9 + 12i - 4) - (9 - 12i - 4)}{13}$$
$$= \frac{24i}{13}$$

14. ARITHMETIC AND GEOMETRIC PROGRESSIONS

1. **(98)** $d = 5$, $n = 20$
$$a_n = a + (n-1)d$$
$$a_n = 3 + (20-1)5$$
$$= 3 + (19)5 = 3 + 95$$
$$= 98$$

2. **(1683)** $d = 3$, $\ell = 99$, $n = 33$
$$S = \frac{n}{2}(d + \ell) = \frac{33}{2}(3 + 99)$$
$$= \frac{33}{2}(102) = 33(51)$$
$$= 1683$$

3. $\left(-\dfrac{1}{32}\right)$ $r = -\dfrac{1}{2}$, $a_1 = 1$, $n = 6$
$$a_n = ar^{n-1} = 1\left(-\frac{1}{2}\right)^5$$
$$= -\frac{1}{32}$$

4. $\left(47\dfrac{5}{8}\right)$ $r = \dfrac{1}{2}$, $a_1 = 24$, $n = 7$
$$S = \frac{a_1 - a_1 r^n}{1 - r} = \frac{24\left[1 - \left(\frac{1}{2}\right)^7\right]}{1 - \frac{1}{2}}$$
$$= \frac{24\left(1 - \frac{1}{128}\right)}{\frac{1}{2}}$$
$$= 48\left(1 - \frac{1}{128}\right)$$
$$= 48 - \frac{3}{8} = 47\frac{5}{8}$$

Solutions to Practice Exercises

5. **(6)** $a_1 = 3, r = \frac{1}{2}$

$$S = \frac{a_1}{1-r} = \frac{3}{1-\frac{1}{2}}$$

$$= 6$$

6. $\left(\frac{5}{33}\right)$ $.151515\ldots = .15 + .0015 + .000015$, etc. $a = .15, r = .01$

$$S = \frac{a}{1-r} = \frac{.15}{1-.01}$$
$$= \frac{.15}{.99} = \frac{5}{33}$$

7. **(108)** $a_1 = 30, d = 2, n = 40$

$$a_1 = a_1 + (n-1)d = 30 + (40-1)2$$
$$= 30 + (39)2$$
$$= 108$$

8. **(2760)**

$$S = \frac{n}{2}(a_1 + a_n)$$

$$S = \frac{40}{2}(30 + 108)$$

$$= 20(138) = 2760$$

15. VECTORS

1. **(100, 37°)**

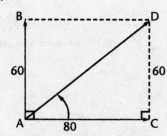

From the figure, right triangle ACD is a 3–4–5 triangle, and so $AD = 100$.

$\tan \angle DAC = \frac{60}{80} = \frac{3}{4} = 0.75$ $\angle DAC = 37°$

2. **(17, S 3° E)**

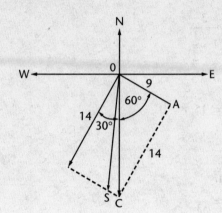

$OC^2 = 9^2 + 14^2 = 81 + 196 = 277$

$OC = \sqrt{277} \approx 17$

Also,

$\tan \angle AOC = \dfrac{14}{9} = 1.5556$

$\angle AOC \approx 57°$

$\angle COS = 60° - 57° = 3°$

Bearing = S 3° E

3. **(130 mph, S 23° E)**

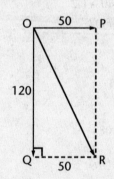

Triangle OQR is a 5–12–13 right triangle, and so $OR = 130$.

$\tan \angle QOR = \dfrac{50}{120} = 0.4167$

$\angle QOR = 23°$

Bearing = S 23° E

4. **(15, 26)**

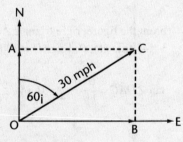

Find the components of the velocity vector. Triangle AOC is a 30–60–90 triangle.

$OA = 15$ and

$OB = 15\sqrt{3} \approx 26$

15 mph northward and 26 mph eastward

5. **(5 mph, 37°)**

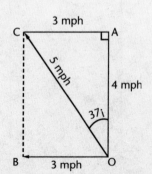

Triangle OAC is a 3–4–5 right triangle. $OC = 5$ mph

$\tan \angle AOC = \dfrac{3}{4} = 0.75$

$\angle AOC = 37°$

SOLUTIONS TO PRACTICE EXERCISES **121**

6. (5 mph, 37° to heading of boat)

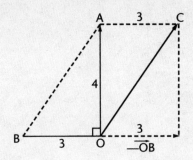

In the figure we are trying to find $\overrightarrow{OA} - \overrightarrow{OB}$. Draw as a vector with magnitude of 3 and opposite in direction to OB. Now find the resultant of $\overrightarrow{OA}$ and $-\overrightarrow{OB}$ from the 3–4–5 right triangle AOC. Then $OC = 5$ and $\tan \angle AOC = \dfrac{3}{4}$ so that $\angle AOC = 37°$.

7. (776 lb)

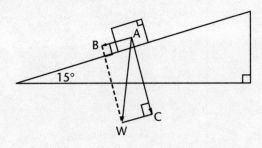

In the figure, $\overrightarrow{AW} = 3{,}000$ lb, $\overrightarrow{AB}$ is the component of $\overrightarrow{AW}$ parallel to the incline and $\overrightarrow{AC}$ perpendicular to the incline. Since the sides of $\angle CAW$ are perpendicular to the sides of the 15° angle, $\angle CAW = 15°$.

Then

$$\sin \angle CAW = \frac{CW}{AW} = \frac{AB}{3{,}000}$$

$$\sin 15° = \frac{AB}{3{,}000}$$

$$0.2588 = \frac{AB}{3{,}000}$$

$$AB = 776$$

8. (26 mph, N 23° E)

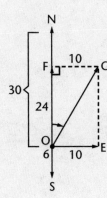

The diagram shows the three forces acting on the ship. The north and south forces have a resultant of 24 heading north.

$$\overrightarrow{OG}^2 = 10^2 + 24^2 = 676$$
$$OG = \sqrt{676} = 26 \text{ mph}$$
$$\tan \angle FOG = \frac{10}{24} = 0.417$$
$$\angle FOG = 23°$$

Bearing is N 23° E.

16. VARIATION

1. **(40)**

$$y = kx$$
$$12 = 3k$$
$$k = 4$$

Substitute in $y = kx$.

$$y = 4x$$

When

$$x = 10$$
$$y = 4(10) = 40$$

2. **(y = 2.4)**

$$y = \frac{k}{x}$$
$$8 = \frac{k}{3} \text{ or } k = 24$$

Substitute in $y = \frac{k}{x}$.

$$y = \frac{24}{x}$$

When

$$x = 10$$
$$y = \frac{24}{10} = 2.4$$

3. $\left(26\frac{2}{3} \text{ lb}\right)$

$$F = kx$$
$$10 = 3k \text{ or } k = \frac{10}{3}$$

Substitute in $F = kx$.

$$F = \frac{10}{3}x$$

When

$$x = 8$$
$$F = \frac{10}{3}(8) = \frac{80}{3} = 26\frac{2}{3} \text{ lb}$$

4. **(8 lb/in.²)**

$$V = \frac{k}{p}$$
$$20 = \frac{k}{12} \text{ or } k = 240$$

Substitute in first equation.

$$V = \frac{240}{p}$$

SOLUTIONS TO PRACTICE EXERCISES

When
$$30 = \frac{240}{p}$$
$$p = 8 \text{ lb/in.}^2$$

5. **(C)**
$$I = \frac{k}{R}$$

Multiply R by 3.
$$I' = \frac{k}{3R} = \frac{1}{3}\frac{k}{R}$$
$$I' = \frac{1}{3}I = \frac{I}{3}$$

I is divided by 3.

6. $(s = 16t^2)$
$$s = kt^2$$
$$1{,}600 = k(10^2)$$
$$1{,}600 = 100k \text{ or } k = 16$$

Substitute in the first equation.
$$s = 16t^2$$

7. **(C)**
$$I = \frac{k}{d^2}$$

Multiply d by 3.
$$I' = \frac{k}{(3d)^2} = \frac{k}{9d^2}$$
$$I' = \frac{1}{9}\frac{k}{d^2} = \frac{1}{9}I = \frac{I}{9}$$

I is divided by 9.

8. $\left(f = \dfrac{1}{\sqrt{L}}\right)$
$$f = \frac{K}{\sqrt{L}}$$
$$1 = \frac{K}{\sqrt{1}} \text{ or } K = 1$$

Substitute in the first equation.
$$f = \frac{1}{\sqrt{L}}$$

9. $\left(V = \dfrac{600T}{P}\right)$ $V = K\dfrac{T}{P}$

$300 = K\dfrac{280}{560} = \dfrac{1}{2}K$

$K = 600$

Substitute in the first equation.

$V = \dfrac{600T}{P}$

10. (B) $F = K\dfrac{mM}{d^2}$

Multiply m by 2 and d by 2.

$F' = K\dfrac{2mM}{(2d)^2}$

$F' = K\dfrac{2mM}{(4d)^2} = \dfrac{1}{2}K\dfrac{mM}{d^2}$

$F' = \dfrac{1}{2}F = \dfrac{F}{2}$

F is divided by 2.

PART FIVE

Four Sample Mathematics Tests Level IC

CONTENTS

Steps to Take after Each Sample Test   127

Sample Test 1: Math Level IC   131

Answer Key   141

Solutions   141

Sample Test 2: Math Level IC   155

Answer Key   164

Solutions   164

Sample Test 3: Math Level IC   179

Answer Key   188

Solutions   188

Sample Test 4: Math Level IC   199

Answer Key   208

Solutions   208

STEPS TO TAKE AFTER EACH SAMPLE TEST

1. Check your answers against the answer key that follows the sample test. Determine your raw score as follows: Count the number of INCORRECT ANSWERS. Take 25% of this number. Subtract 25% of the incorrect answers from the number of CORRECT ANSWERS. For example, let us assume that out of the 50 questions you have gotten 39 correct and 12 incorrect and have omitted 2. 25% of 12 = 3, so subtract 3 from 39. 39 − 3 = 36. Your raw score on this sample test is 36. (Reported scores are scaled scores, on a scale of 200–800.)

2. Determine your unofficial percentile ranking by consulting the table below.

PERCENTILE RANKING TABLE
(*Unofficial*)

Approximate Percentile Ranking	Raw Score	Approximate Percentile Ranking	Raw Score	Approximate Percentile Ranking	Raw Score
99	50	77–78	39	57–58	29
97–98	49	75–76	38	55–56	28
95–96	48	73–74	37	53–54	27
93–94	47	71–72	36	51–52	26
91–92	46	69–70	35	49–50	25
89–90	45	67–68	34	47–48	24
87–88	44	65–66	33	45–46	23
85–86	43	63–64	32	43–44	22
83–84	42	61–62	31	41–42	21
81–82	41	59–60	30	0–40	Under 21
79–80	40				

3. Carefully go over the solutions to all the questions that you answered incorrectly. Pinpoint the areas in which you show weakness. Use the following Diagnostic Checklist to establish the areas that require the greatest application on your part. One check mark after an item means that you are moderately weak; two check marks means that you are seriously weak.

128 SAT II: MATH

DIAGNOSTIC CHECKLIST

Area of Weakness	Check Below
ALGEBRA	
APPROXIMATION	
EQUATIONS AND VARIATION	
EUCLIDEAN GEOMETRY	
EXPONENTS	
FORMULAS AND LINEAR EQUATIONS	
FRACTIONS	
FUNCTIONS	
GEOMETRY AND VECTORS	
GRAPHS AND COORDINATE GEOMETRY	
INEQUALITIES	
LOGARITHMS	
LOGIC AND PROOF	
NUMBER THEORY	
PROBABILITY	
SEQUENCES AND LIMITS	
SETS	
TRIGONOMETRY	
VERBAL PROBLEMS	

4. Eliminate the weaknesses that you just indicated in the preceding Diagnostic Checklist. Do the following:

    a) Get the assistance of a person who is knowledgeable in those areas of Mathematics in which your weaknesses have been exposed.

    b) Refer to textbooks and study to eliminate your weaknesses.

5. Go on to another sample test. Again place yourself under examination conditions. After each sample test repeat the scoring, diagnostic and self-improvement procedures. If you diligently and systematically follow this plan, you should do better on your last sample test than you did on the first one, and you should be well prepared for your Mathematics Subject Test.

Sample Test 1
Answer Sheet

Math Level IC

1. Ⓐ Ⓑ Ⓒ Ⓓ Ⓔ
2. Ⓐ Ⓑ Ⓒ Ⓓ Ⓔ
3. Ⓐ Ⓑ Ⓒ Ⓓ Ⓔ
4. Ⓐ Ⓑ Ⓒ Ⓓ Ⓔ
5. Ⓐ Ⓑ Ⓒ Ⓓ Ⓔ
6. Ⓐ Ⓑ Ⓒ Ⓓ Ⓔ
7. Ⓐ Ⓑ Ⓒ Ⓓ Ⓔ
8. Ⓐ Ⓑ Ⓒ Ⓓ Ⓔ
9. Ⓐ Ⓑ Ⓒ Ⓓ Ⓔ
10. Ⓐ Ⓑ Ⓒ Ⓓ Ⓔ
11. Ⓐ Ⓑ Ⓒ Ⓓ Ⓔ
12. Ⓐ Ⓑ Ⓒ Ⓓ Ⓔ
13. Ⓐ Ⓑ Ⓒ Ⓓ Ⓔ
14. Ⓐ Ⓑ Ⓒ Ⓓ Ⓔ
15. Ⓐ Ⓑ Ⓒ Ⓓ Ⓔ
16. Ⓐ Ⓑ Ⓒ Ⓓ Ⓔ
17. Ⓐ Ⓑ Ⓒ Ⓓ Ⓔ
18. Ⓐ Ⓑ Ⓒ Ⓓ Ⓔ
19. Ⓐ Ⓑ Ⓒ Ⓓ Ⓔ
20. Ⓐ Ⓑ Ⓒ Ⓓ Ⓔ
21. Ⓐ Ⓑ Ⓒ Ⓓ Ⓔ
22. Ⓐ Ⓑ Ⓒ Ⓓ Ⓔ
23. Ⓐ Ⓑ Ⓒ Ⓓ Ⓔ
24. Ⓐ Ⓑ Ⓒ Ⓓ Ⓔ
25. Ⓐ Ⓑ Ⓒ Ⓓ Ⓔ
26. Ⓐ Ⓑ Ⓒ Ⓓ Ⓔ
27. Ⓐ Ⓑ Ⓒ Ⓓ Ⓔ
28. Ⓐ Ⓑ Ⓒ Ⓓ Ⓔ
29. Ⓐ Ⓑ Ⓒ Ⓓ Ⓔ
30. Ⓐ Ⓑ Ⓒ Ⓓ Ⓔ
31. Ⓐ Ⓑ Ⓒ Ⓓ Ⓔ
32. Ⓐ Ⓑ Ⓒ Ⓓ Ⓔ
33. Ⓐ Ⓑ Ⓒ Ⓓ Ⓔ
34. Ⓐ Ⓑ Ⓒ Ⓓ Ⓔ
35. Ⓐ Ⓑ Ⓒ Ⓓ Ⓔ
36. Ⓐ Ⓑ Ⓒ Ⓓ Ⓔ
37. Ⓐ Ⓑ Ⓒ Ⓓ Ⓔ
38. Ⓐ Ⓑ Ⓒ Ⓓ Ⓔ
39. Ⓐ Ⓑ Ⓒ Ⓓ Ⓔ
40. Ⓐ Ⓑ Ⓒ Ⓓ Ⓔ
41. Ⓐ Ⓑ Ⓒ Ⓓ Ⓔ
42. Ⓐ Ⓑ Ⓒ Ⓓ Ⓔ
43. Ⓐ Ⓑ Ⓒ Ⓓ Ⓔ
44. Ⓐ Ⓑ Ⓒ Ⓓ Ⓔ
45. Ⓐ Ⓑ Ⓒ Ⓓ Ⓔ
46. Ⓐ Ⓑ Ⓒ Ⓓ Ⓔ
47. Ⓐ Ⓑ Ⓒ Ⓓ Ⓔ
48. Ⓐ Ⓑ Ⓒ Ⓓ Ⓔ
49. Ⓐ Ⓑ Ⓒ Ⓓ Ⓔ
50. Ⓐ Ⓑ Ⓒ Ⓓ Ⓔ

Directions: For each question in the sample test, select the best of the answer choices and blacken the corresponding space on this answer sheet.

Please note: (a) You will need to use a calculator in order to answer some, though not all, of the questions in this test. As you look at each question, you must decide whether or not you need a calculator for the specific question. A four-function calculator is not sufficient; your calculator must be at least a scientific calculator. Calculators that can display graphs and programmable calculators are also permitted.

(b) The only angle measure used on the Level IC test is degree measure. Your calculator should be set to degree mode.

(c) All figures are accurately drawn and are intended to supply useful information for solving the problems that they accompany. Figures are drawn to scale UNLESS it is specifically stated that a figure is not drawn to scale. Unless otherwise indicated, all figures lie in a plane.

(d) The domain of any function f is assumed to be the set of all real numbers x for which $f(x)$ is a real number except when this is specified not to be the case.

(e) Use the reference data below as needed.

REFERENCE DATA

Solid	Volume	Other		
Right circular cone	$V = \frac{1}{3}\pi r^2 h$	$S = cl$	V = volume r = radius h = height	S = lateral area c = circumference of base l = slant height
Sphere	$V = \frac{4}{3}\pi r^3$	$S = 4\pi r^2$	V = volume r = radius S = surface area	
Pyramid	$V = \frac{1}{3}Bh$		V = volume B = area of base h = height	

Sample Test 1

MATH LEVEL IC

50 Questions • Time—60 Minutes

1. Which of the following illustrates a distributive principle?

    (A) $5 + 4 = 4 + 5$
    (B) $(3 + 4) + 5 = 3 + (4 + 5)$
    (C) $(6 \cdot 2) + 4 = (2 \cdot 6) + 4$
    (D) $6 \cdot (2 \cdot 4) = (4 \cdot 2) \cdot 6$
    (E) $6 \cdot (2 + 4) = 6 \cdot 2 + 6 \cdot 4$

2.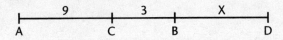

    In the figure, $AC = 9$, $BC = 3$, and D is 3 times as far from A as from B. What is BD?

    (A) 6   (B) 9   (C) 12   (D) 15   (E) 18

3. If n is a positive integer, which of the following is always odd?

    (A) $19n + 6$ (B) $19n + 5$ (C) $19n^2 + 5$ (D) $18n + 4$
    (E) $18n + 5$

4. Let R be the set of all numbers r such that $-5 < r < 8$. Let S be the set of all numbers s such that $3 < s < 10$. The intersection T of R and S is the set of all numbers t such that

    (A) $-5 < t < 3$
    (B) $-5 < t < 8$
    (C) $0 < t < 8$
    (D) $3 < t < 8$
    (E) $8 < t < 10$

5. If $b > 1$ and $b^y = 1.5$, then $b^{-2y} =$

    (A) -3.0 (B) -2.25 (C) $-\frac{1}{2.25}$ (D) $\frac{1}{2.25}$ (E) $\frac{1}{3.0}$

6. $(y-2)(y+7)^2 < 0$, if and only if

   (A) $y < 2$
   (B) $-7 < y < 2$
   (C) $y > -7$
   (D) $y < 2$ and $y \neq -7$
   (E) $2 < y < 7$ and $y > 7$

7. A computer is programmed to add 3 to the number N, multiply the result by 3, subtract 3, and divide this result by 3. The computer answer will be

   (A) $N + 1$
   (B) $N + 2$
   (C) N
   (D) $N - 2$
   (E) $N + \frac{1}{3}$

8. $\dfrac{2x-4}{4x+20} \cdot \dfrac{x^2-25}{x^2-7x+10} = (x \neq 5)$

   (A) $\dfrac{1}{4}$ (B) $\dfrac{1}{2}$ (C) $\dfrac{2}{x-5}$ (D) $\dfrac{2}{x+5}$ (E) none of these

9. If $S = \dfrac{rL-a}{r-1}$, $r =$

   (A) $\dfrac{a-S}{S-L}$ (B) $\dfrac{1}{S-L}$ (C) $\dfrac{S-a}{S-L}$ (D) 1 (E) $\dfrac{S-a}{L-S}$

10. $\dfrac{1 - \frac{9}{y^2}}{1 - \frac{3}{y}} - \dfrac{3}{y} = (y \neq 0)$

    (A) $\dfrac{y-3}{y}$ (B) $\dfrac{y+3}{y}$ (C) 3 (D) 1 (E) $3y - 1$

11. If I varies inversely as d^2 and $I = 20$ when $d = 3$, what is I when $d = 10$?

    (A) 6 (B) $66\frac{2}{3}$ (C) 18 (D) 1.8 (E) 12

12. If y is the measure of an acute angle such that $\sin y = \dfrac{a}{5}$, $\tan y =$

    (A) $\dfrac{\sqrt{25-a^2}}{5}$ (B) $\dfrac{a}{\sqrt{25-a^2}}$ (C) $\dfrac{5}{\sqrt{25-a^2}}$ (D) $\dfrac{a}{5-a}$

    (E) $\dfrac{a}{\sqrt{25+a^2}}$

DO YOUR FIGURING HERE.

13. How many degrees between the hands of a clock at 3:40?

    (A) 150°   (B) 145°   (C) 140°   (D) 135°   (E) 130°

14. The legs of a right triangle are in the ratio of 1:2. If the area of the triangle is 25, what is the hypotenuse?

    (A) $5\sqrt{5}$ (B) $5\sqrt{3}$ (C) 10   (D) $10\sqrt{3}$ (E) $10\sqrt{5}$

15. If $x = 1 + \sqrt{2}$, then $x^2 - 2x + 1$ equals

    (A) $1 + \sqrt{2}$ (B) $\sqrt{2-1}$ (C) 2   (D) $\sqrt{2}$ (E) $2 + \sqrt{2}$

16. A point is 17 in. from the center of a circle of radius 8 in. The length of the tangent from the point to the circle is

    (A) $\sqrt{353}$ (B) 15   (C) 9   (D) $9\sqrt{3}$ (E) $15\sqrt{2}$

17. In the formula $f = \frac{C}{L}$, if $C = 3 \times 10^{10}$ and $L = 6 \times 10^{-5}$, $f =$

    (A) 2×10^{15} (B) 2×10^5 (C) 5×10^{14} (D) 2×10^{14}
    (E) 5×10^{15}

18. What is the slope of the line $\sqrt{14}\ x - 3y = \sqrt[3]{7}$?

    (A) 1.15
    (B) 1.25
    (C) 1.35
    (D) 1.45
    (E) 1.55

19. How many numbers in the set $\{-8, -5, 0, 10, 20\}$ satisfy the condition $|x - 5| < 11$?

    (A) none   (B) one   (C) two   (D) three   (E) four

20. The graph of $x^2 - \sqrt{5}x - 2$ has its minimum value at which value of x?

    (A) .83
    (B) 1.12
    (C) 1.21
    (D) 1.35
    (E) 2.47

21. In △PQR, if the measure of ∠Q is 50° and the measure of ∠P is $p°$, and if PQ is longer than PR, then

(A) $0 < p < 40$
(B) $0 < p < 80$
(C) $40 < p < 80$
(D) $80 < p < 90$
(E) $80 < p < 130$

22. Three parallel lines are cut by three non-parallel lines. What is the maximum number of points of intersection of all six lines?

(A) 10   (B) 11   (C) 12   (D) 13   (E) 14

23. In figure 23, ∠QSR = 30° in circle 0. What is the value of angle QPR?

(A) 10
(B) 15
(C) 20
(D) 25
(E) cannot be determined from the information given

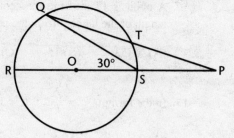

Fig. 23

24. If $f(x) = x^3 - 3$ and $g(x) = 7x - 5$, what is the value of $f(g(3.9))$?

(A) 389.23
(B) 938.32
(C) 4261.48
(D) 11086.57
(E) 14257.91

25. For what value(s) of y on the curve shown in figure 25 does $y = 4x$?

(A) no value
(B) +4 only
(C) +3 only
(D) −5 only
(E) +4 and −12

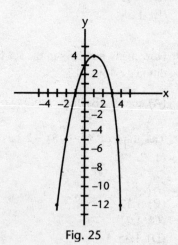

Fig. 25

26. In figure 26, *RT* is a diameter of the semicircle. If *RS* = 2 and *ST* = 3, then the area of the semicircle is

(A) $\dfrac{13\pi}{2}$

(B) $\dfrac{13\pi}{4}$

(C) $\dfrac{13\pi}{6}$

(D) $\dfrac{13\pi}{8}$

(E) cannot be determined from the information given

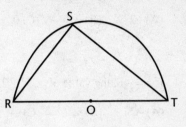

Fig. 26

27. A triangle with vertices (0, 0), (4, 3), and (− 3, 4) belongs to which of the following classes?

  I—Scalene Triangles
  II—Isosceles Triangles
  III—Right Triangles
  IV—Equilateral Triangles

(A) none
(B) I only
(C) II and III only
(D) IV only
(E) III only

28. The equation of the graph in figure 28 is

(A) $y = |x|$
(B) $y = x$
(C) $y = -x$
(D) $y = 2x$
(E) $y = x^2$

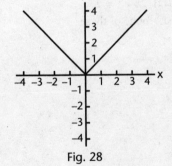

Fig. 28

29. What is the length of the line segment joining the points *N* (7, −2) and *J* (−2, 7)?

(A) 3.16 (B) 9.83 (C) 10.00 (D) 11.42
(E) 12.73

30. If the graph of the equation $x + y - 8 + 4k = 0$ passes through the origin, the value of k is

    (A) –2   (B) 2   (C) 0   (D) 1   (E) –1

31. If $x = -8$, the value of $x^{2/3} + 2x^0$ is

    (A) –2   (B) 4   (C) –4   (D) 6   (E) –6

32. If $2^{3x+10} = \left(\frac{1}{4}\right)^x$, $x =$

    (A) 0   (B) 1   (C) –1   (D) 2   (E) –2

33.

x	2	4	7	11
y	3	7	13	21

The equation expressing the relationship between x and y in the above table is

(A) $y = 2x + 1$
(B) $y = x + 2$
(C) $y = 2x - 1$
(D) $2x + y = 7$
(E) none of these

DO YOUR FIGURING HERE.

34. The graph of the equation $x^2 - 2y^2 = 8$ is

   (A) a circle
   (B) an ellipse
   (C) a hyperbola
   (D) a parabola
   (E) two straight lines

35. The fraction $\dfrac{\sqrt{3} - \sqrt{2}}{\sqrt{2}}$ is equal to

   (A) $\sqrt{3}$
   (B) $\dfrac{\sqrt{3} - 2}{2}$
   (C) $\sqrt{6}$
   (D) $\dfrac{\sqrt{6} - 2}{2}$
   (E) $\sqrt{3} - 1$

36. For what values of K does the equation $Kx^2 - 4x + K = 0$ have real roots?

   (A) $+2$ and -3
   (B) $-2 \leq K \leq 2$
   (C) $K \leq 2$
   (D) $K \geq -2$
   (E) $-4 \leq K \leq 4$

37. The radiator of a car contains 10 quarts of a 20% solution of alcohol. If 2 quarts of water are added, what percent of the resulting solution is alcohol?

   (A) 18% (B) $16\tfrac{2}{3}$% (C) $15\tfrac{1}{4}$% (D) 14% (E) $12\tfrac{1}{2}$%

38. Express the infinite decimal .212121 . . . as a common fraction.

   (A) $\dfrac{21}{100}$ (B) $\dfrac{23}{99}$ (C) $\dfrac{7}{100}$ (D) $\dfrac{7}{99}$ (E) $\dfrac{7}{33}$

GO ON TO THE NEXT PAGE

39. *PQ* and *PT* are tangent to circle *O*. If angle *P* = 70°, how many degrees in minor arc *QT*?

    (A) 140
    (B) 125
    (C) 120
    (D) 110
    (E) 100

DO YOUR FIGURING HERE.

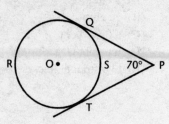

Fig. 39

40. A cubic foot of water is poured into a rectangular aquarium with base 15 in. by 18 in. To what height in inches does the water rise?

    (A) $6\frac{2}{5}$ (B) 6   (C) $5\frac{3}{4}$ (D) $5\frac{1}{2}$ (E) 5

41. A car drives a distance of *d* miles at 30 mph and returns at 60 mph. What is its average rate for the round trip?

    (A) 45 mph   (B) 43 mph   (C) 40 mph   (D) $\frac{2d}{35}$ mph

    (E) $\frac{35}{2d}$ mph

42. If $y = \sqrt{7}x^2 + \sqrt{5}x + \sqrt{3}$, what is the sum of the roots?

    (A) 1.53
    (B) 1.18
    (C) −.65
    (D) −.77
    (E) −.85

43. A circle is inscribed in a triangle with sides 9, 12, and 15. The radius of the circle is

    (A) 2   (B) 3   (C) 3.5   (D) 4   (E) 4.6

44. The interior angles of a regular polygon are each 165°. How many sides does the polygon have?

    (A) 17   (B) 20   (C) 22   (D) 24   (E) 28

45. Find the root(s) of the equation $y + \sqrt{y+5} = 7$.

    (A) 11 (B) 4 (C) 4 and 11 (D) ±4 (E) none of these

46. Which of the following is the equation of a line perpendicular to $\frac{x}{.47} + \frac{y}{.53} = 1$ and passing through the point $\left(\frac{1}{2}, -\frac{1}{2}\right)$

    (A) $\left(y - \frac{1}{2}\right) = -1.13\left(x - \frac{1}{2}\right)$

    (B) $\left(x - \frac{1}{2}\right) = .89\left(y + \frac{1}{2}\right)$

    (C) $\left(y - \frac{1}{2}\right) = -1.13\left(x + \frac{1}{2}\right)$

    (D) $\left(y + \frac{1}{2}\right) = -1.13\left(x - \frac{1}{2}\right)$

    (E) $\left(y + \frac{1}{2}\right) = .89\left(x - \frac{1}{2}\right)$

47. In how many points do the graphs of the equations $x^2 + y^2 = 25$ and $y^2 = 4x$ intersect?

    (A) 0 (B) 1 (C) 2 (D) 3 (E) 4

48. In right triangle ABC, $DE \perp BC$. If $AB = 6$, $AC = 8$, $BC = 10$ and $DE = 4$, find EC.

    (A) $5\frac{1}{3}$

    (B) $6\frac{2}{3}$

    (C) 5

    (D) 6

    (E) $4\frac{3}{4}$

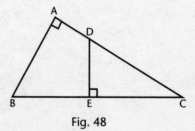

Fig. 48

49. A man can do a job in h hours alone and his son can do it in $2h$ hours alone. Together, how many hours will it take them to do the job?

    (A) $3h$ (B) $\frac{h}{3}$ (C) $\frac{3h}{2}$ (D) $\frac{2h}{3}$ (E) $\frac{h}{2}$

50. The diagonals of a parallelogram divide the figure into four triangles which are

   (A) congruent
   (B) similar
   (C) equal in area
   (D) isosceles
   (E) none of these

DO YOUR FIGURING HERE.

STOP

Sample Test 1
Answer Key

Math Level IC

1. E	11. D	21. B	31. D	41. C
2. A	12. B	22. C	32. E	42. E
3. E	13. E	23. E	33. C	43. B
4. D	14. A	24. D	34. C	44. D
5. D	15. C	25. E	35. D	45. B
6. D	16. B	26. D	36. B	46. E
7. B	17. C	27. C	37. B	47. C
8. B	18. B	28. A	38. E	48. A
9. C	19. D	29. E	39. D	49. D
10. D	20. B	30. B	40. A	50. C

Solutions

1. **(E)** The distributive principle refers to the product of a single quantity and sum of quantities; that is, $a(b + c) = ab + ac$.

2. **(A)**

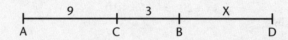

Let $BD = x$. Then $AD = 3BD$ or $12 + x = 3x$.
Subtract x from both sides.

$$12 = 2x$$
$$x = 6$$

3. **(E)** Examine each choice in turn.

   (A) $19n + 6$: If n is even, $19n$ is even, and the sum of two even numbers is even.
   (B) $19n + 5$: If n is odd, $19n$ is odd, and the sum of two odd numbers is even.
   (C) $19n^2 + 5$: If n is odd, n^2 is odd, $19n^2$ is odd, and the sum of two odd numbers is even.
   (D) $18n + 4$: $19n$ is always even, and the sum of two even numbers is even.
   (E) $18n + 5$: $18n$ is always even, and the sum of an even and an odd number is odd.

4. **(D)** Put both sets on a number line and determine their intersection.

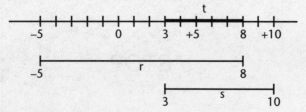

The heavy line is the intersection of the sets, $3 < t < 8$.

5. **(D)**
$$b^{-2y} = \frac{1}{b^{2y}} = \frac{1}{\left(b^y\right)^2}$$

These equalities follow from the laws of exponents. Substitute $b^y = 1.5$.

$$b^{-2y} = \frac{1}{(1.5)^2} = \frac{1}{2.25}$$

6. **(D)**
$$(y-2)(y+7)^2 < 0$$

$(y+7)^2$ is always a positive quantity when $y \neq -7$. $(y-2)$ must then be a negative quantity to make the above product negative.
$$y-2<0$$
$$y<2$$

7. **(B)**
$$\frac{3(N+3)-3}{3} = \frac{3N+9-3}{3}$$
$$= \frac{3N+6}{3}$$
$$= N+2$$

8. **(B)**
$$\frac{2x-4}{4x+20} \cdot \frac{x^2-25}{x^2-7x+10}$$

Factor wherever possible.

$$\frac{\cancel{2}(\cancel{x-2})}{\cancel{2}4(x+5)} \cdot \frac{(\cancel{x-5})(x+5)}{(\cancel{x-2})(\cancel{x-5})} = \frac{1}{2}$$

9. **(C)**
$$S = \frac{rL-a}{r-1}$$

Multiply both sides by $(r-1)$.
$$S(r-1) = rL-a$$
$$Sr-S = rL-a$$

Add S and $-rL$ to both sides.
$$Sr-rL = S-a$$
$$r(S-L) = S-a$$

Divide both sides by $S-L$.
$$r = \frac{S-a}{S-L}$$

SOLUTIONS   143

10. **(D)**
$$\frac{1-\frac{9}{y^2}}{1-\frac{3}{y}}-\frac{3}{y}$$

$$=\frac{\left(1+\frac{3}{y}\right)\left(1-\frac{3}{y}\right)}{\left(1-\frac{3}{y}\right)}-\frac{3}{y}$$

$$=1+\frac{3}{y}-\frac{3}{y}=1$$

11. **(D)** $I=\frac{K}{d^2}$ where K is the constant of proportionality. To determine K, substitute $I=20$ and $d=3$.

$$20=\frac{K}{9} \text{ or } K=180$$

The formula then becomes $I=\frac{180}{d^2}$.

Substitute $d=10$ in the formula.

$$I=\frac{180}{100}$$

$$I=1.8$$

12. **(B)**

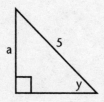

Construct a right triangle with a hypotenuse 5 and leg a opposite $\angle y$.

$$\sin y = \frac{a}{5}$$

By the Pythagorean theorem, the leg adjacent to $\angle y$ becomes $\sqrt{25-a^2}$.

$$\tan \angle y = \frac{\text{opposite leg}}{\text{adjacent leg}} = \frac{a}{\sqrt{25-a^2}}$$

13. **(E)**

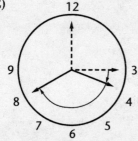

Consider the position of the hands at 3 o'clock. The large hand is at 12 and the small hand at 3. At 3:40 the large hand is at 8 and the small hand has moved $\frac{2}{3}$ of the distance between the 3 and 4. Since there are 30° between the 3 and the 4, the small hand has moved $\frac{2}{3} \times 30° = 20°$. Between the 3 and the 8 there are $5 \times 30° = 150°$ of arc.

Therefore at 3:40, the angle between the hands is $150° - 20° = 130°$.

14. **(A)** Designate the legs of the right triangle by x and $2x$. The area of the triangle is then

$$A = \frac{1}{2} \cdot x \cdot 2x = 25$$

$$\frac{1}{2} \cdot 2x^2 = 25$$

$$x^2 = 25$$

$$x = 5 \text{ and } 2x = 10$$

If the legs are 5 and 10, the hypotenuse y is

$$y^2 = 5^2 + 10^2 = 125$$

$$y = \sqrt{125} = \sqrt{25 \cdot 5} = 5\sqrt{5}$$

15. **(C)** Substitute $x = 1 + \sqrt{2}$ in the expression $x^2 - 2x + 1$.

$$\left(1+\sqrt{2}\right)^2 - 2\left(1+\sqrt{2}\right) + 1$$

$$= 1 + 2\sqrt{2} + 2 - 2 - 2\sqrt{2} + 1 = 2$$

or

$$x^2 - 2x + 1 = (x-1)^2 = \left(1 + \sqrt{2} - 1\right)^2 = 2$$

16. **(B)**

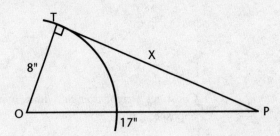

Let the tangent $PT = x$; then $OP = 17$ and $OT = 8$.

The radius $OT \perp PT$ so that OPT is a right $\triangle$.

$$x^2 + 8^2 = 17^2 \text{ by the Pythagorean Theorem}$$

$$x^2 + 64 = 289$$

$$x^2 = 225$$

$$x = 15$$

17. **(C)**

$$f = \frac{C}{L}$$

Substitute,

$$f = \frac{3 \cdot 10^{10}}{6 \cdot 10^{-5}}$$

When we divide powers of the same base, we subtract exponents.

$$f = \frac{1}{2} \times 10^{15}$$

$$= .5 \times 10^{15}$$

$$= 5 \times 10^{14}$$

18. **(B)** $\sqrt{14}x - 3y = \sqrt[3]{7} \Rightarrow 3y = \sqrt{14}x - \sqrt[3]{7}$

$$y = \frac{\sqrt{14}}{3}x - \frac{\sqrt[3]{7}}{3}$$

$$\text{slope} = \frac{\sqrt{14}}{3} = 1.247$$

19. **(D)** $|x - 5| < 11$ is equivalent to

$x - 5 < 11$ when $x - 5 > 0$ or $x > 5$.
thus $x < 16$ when $x > 5$ or $5 < x < 16$.

Only the value $x = 10$ in the given set is in this interval.

$|x - 5|$ is equivalent to $-(x - 5)$ when $x - 5 < 0$ or $x < 5$.
Solving the inequality $5 - x < 11$, we get $-x < 6$ or $x > -6$ when $x < 5$.

$$\text{Or } -6 < x < 5.$$

The values $x = -5$ and $x = 0$ in the given set lie in this interval. Hence, there are three such values.

An alternate method of solution would be to list each of the 5 values of the given set in the inequality.

$|-8 - 5| = |-13| = 13$, which is not less than 11, etc.

20. **(B)** The x value of the minimum point is

$$x = \frac{-b}{2a} = \frac{-\left(-\sqrt{5}\right)}{2(1)} = \frac{\sqrt{5}}{2} = 1.12$$

21. **(B)**

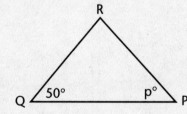

If $PQ > PR$, then $\angle R > \angle Q$, since the larger angle lies opposite the longer side. If $\angle R > 50°$, and $\angle Q = 50°$, then p is less than 80, since there are 180° in the sum of the angles of a triangle. However, $\angle R$ may have any value less than 130°, in which case p must be greater than but not equal to 0.

$$0 < p < 80$$

22. **(C)** 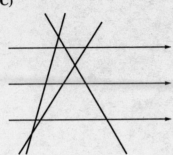 The three parallel lines intersect each of the three non-parallel lines in 3 points, making a total of 9. The 3 non-parallel lines form a triangle, giving us 3 more points of intersection at the vertices. Hence, there are a total of 12.

23. **(E)** In order to determine angle P from the figure given in the problem, we have to know both arcs intercepted on the circle by PQ and PR. Arc QR is apparently 60°, but there is no way of determining arc ST. Hence P cannot be determined from the given information.

24. **(D)** $g(3.9) = 7(3.9) - 5$
    $= 22.3$
    $f(22.3) = (22.3)^3 - 3$
    $= 11086.567$

25. **(E)** Draw the line graph of $y = 4x$ on the same set of axes. This line passes through the origin and has a slope of 4. It thus intersects the curve in (1,4) and (−3, −12). Thus the desired values of y are +4 and −12.

26. **(D)** Angle S is a right angle since it is inscribed in a semicircle. Thus $\triangle RST$ is a right $\triangle$ with RT the hypotenuse. By the Pythagorean Theorem

$$\overline{RT}^2 = \overline{RS}^2 + \overline{ST}^2$$
$$\overline{RT}^2 = 2^2 + 3^2$$
$$\overline{RT}^2 = 4 + 9 = 13$$
$$RT = \sqrt{13}$$

radius OT is then $= \frac{1}{2}\sqrt{13}$

area of semicircle $= \frac{1}{2}\pi r^2$
$$= \frac{1}{2}\pi\left(\frac{1}{2}\sqrt{13}\right)^2 = \frac{1}{2}\pi \cdot \frac{13}{4}$$
$$= \frac{13\pi}{8}$$

27. **(C)**

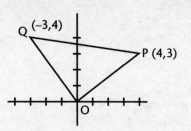

slope of $OP = \frac{3}{4}$

slope of $OQ = -\frac{4}{3}$

Since slopes are negative reciprocals, $OP \perp OQ$

Also $OP = \sqrt{3^2 + 4^2} = \sqrt{9+16} = \sqrt{25} = 5$

$OQ = \sqrt{(-3)^2 + 4^2} = \sqrt{9+16} = 5$

$OP = OQ$ and the triangle is right, isosceles.

28. **(A)** The line in the first quadrant is the graph of $y = x$ for $x \geq 0$.
The line in the second quadrant is the graph of $y = -x$ for $x \leq 0$.
$y = |x|$ means $y = x$ for $x \geq 0$ and $y = -x$ for $x \leq 0$.
Hence, the equation is $y = |x|$.

29. **(E)**

$$d = \sqrt{(x_2 - x_1)^2 + (y_2 - y_1)^2}$$
$$= \sqrt{(-2-7)^2 + (7-(-2))^2}$$
$$= \sqrt{81+81}$$
$$= \sqrt{162}$$
$$= 12.73$$

30. **(B)** If the graph of the equation passes through the origin, the values $x = 0$, $y = 0$ must satisfy the equation.

Substitute.

$$0 + 0 - 8 + 4K = 0$$
$$4K = 8 \text{ or } K = 2.$$

31. **(D)** Substitute -8 for x in $x^{2/3} + 2x^0$.

$$(-8)^{2/3} + 2 \cdot 8^0$$
$$= \sqrt[3]{(-8)^2} + 2 \cdot 1$$
$$= \sqrt[3]{64} + 2 = 4 + 2 = 6$$

32. **(E)**

$$2^{3x+10} = \left(\frac{1}{4}\right)^x = \left(2^{-2}\right)^x$$

$$2^{3x+10} = 2^{-2x}$$

Set the exponents equal.

$$3x + 10 = -2x$$
$$5x = -10$$
$$x = -2$$

33. **(C)** Make a table of Δx and Δy (change in x and y).

Δx	2	3	4
Δy	4	6	8

$$\frac{\Delta y}{\Delta x} = \frac{4}{2} = \frac{6}{3} = \frac{8}{4} = 2$$

Since the slope is constant, the graph is a straight line of the form $y = 2x + b$. Substituting $x = 2$, $y = 3$ we see $b = -1$. Hence, the equation is $y = 2x - 1$.

34. **(C)** Divide both sides by 8.

$$\frac{x^2}{8} - \frac{y^2}{4} = 1$$

This now resembles the standard form

$$\frac{x^2}{a^2} - \frac{y^2}{b^2} = 1$$

which is the equation of a hyperbola.

35. **(D)** Rationalize the denominator by multiplying numerator and denominator by $\sqrt{2}$.

$$\frac{\sqrt{3} - \sqrt{2}}{\sqrt{2}} \cdot \frac{\sqrt{2}}{\sqrt{2}} = \frac{\sqrt{6} - 2}{2}$$

36. **(B)** For the equation $Kx^2 - 4x + K = 0$ to have real roots, its discriminant must be ≥ 0.

$$16 - 4K^2 \geq 0$$
$$16 \geq 4K^2$$
$$4 \geq K^2$$

This is equivalent to

$$|K| \leq 2 \text{ or } -2 \leq K \leq 2$$

37. **(B)** In the original solution there are $.20 \times 10 = 2$ quarts of alcohol. After 2 quarts of water are added, the resulting solution has the same 2 quarts of alcohol in 12 quarts of solution.

$$\frac{2}{12} = \frac{1}{6} = 16\frac{2}{3}\%$$

38. **(E)** Write $.212121\ldots$ as the sum of the terms of an infinite geometric progression.

$$S = .21 + .0021 + .000021 + \cdots$$

$$a = .21 \text{ and } r = .01$$

$$S = \frac{a}{1-r} = \frac{.21}{1-.01} = \frac{.21}{.99}$$

$$S = \frac{21}{99} = \frac{7}{33}$$

39. **(D)** $\angle P \stackrel{\circ}{=} \frac{1}{2}(\overset{\frown}{QRT} - \overset{\frown}{QST})$

Let $QST = x°$
Then $QRT = 360 - x°$
Substitute in the first equation.

$$70 = \frac{1}{2}(360 - x - x)$$

Combine terms and multiply by 2.

$$140 = 360 - 2x$$
$$2x = 360 - 140$$
$$2x = 220$$
$$x = 110$$

40. **(A)** Assume the water rises to a height of x inches.
Then $15 \cdot 18 \cdot x = 12 \cdot 12 \cdot 12$ (1 cu ft).
Divide both sides by 18.

$$15x = 96$$
$$x = 6\frac{2}{5}$$

41. **(C)**

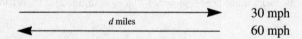
d miles   30 mph
   60 mph

$$\text{Average rate} = \frac{\text{total distance}}{\text{total time}}$$

The time going is $\frac{d}{30}$ and the time coming is $\frac{d}{60}$.
The total time is

$$\frac{d}{30} + \frac{d}{60} = \frac{2d}{60} + \frac{d}{60} = \frac{3d}{60} = \frac{d}{20}$$

$$\text{Average rate} = \frac{2d}{d/20} = \frac{40d}{d} = .40 \text{ mph}$$

42. **(E)**   Sum of roots $= \frac{-b}{a}$

$$= \frac{-\sqrt{5}}{\sqrt{7}}$$

$$= -.845$$

43. (B)

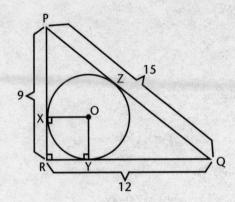

Since $15^2 = 9^2 + 12^2 = 81 + 144 = 225$, $\triangle PQR$ is a right triangle with a right angle at R. $OX = OY$ = radius of the inscribed circle, and since $OXRY$ is a square the radius also equals RX or RY.

Let $RX = RY = r$.
Then $PX = 9 - r = PZ$ and $QY = 12 - r = QZ$
Since $PZ + QZ = 15$
$9 - r + 12 - r = 15$
$\quad\quad 21 - 2r = 15$
$\quad\quad\quad\quad r = 3$

44. (D) If each interior angle is 165°, each exterior angle is 180° − 165° = 15°. Since the sum of the exterior angles is 360°, there are $\frac{360}{15}$ = 24 exterior angles and therefore 24 sides.

45. (B) $\quad\quad\quad\quad\quad\quad\quad y + \sqrt{y+5} = 7$

Subtract y from both sides.

$$\sqrt{y+5} = 7 - y$$

Square both sides.

$$y + 5 = (7 - y)^2 = 49 - 14y + y^2$$

Subtract $(y + 5)$ from both sides,

$$y^2 - 15y + 44 = 0$$
$$(y - 11)(y - 4) = 0$$
$$y = 11 \text{ or } y = 4$$

Substituting in the original equation, we see that only $y = 4$ checks.

46. **(E)** First find the slope of the given line

$$\frac{x}{.47} + \frac{y}{.53} = 1$$

Multiply the entire equation by $(.47)(.53)$
$.53x + .47y = (.47)(.53)$
$y = -1.128x + .53$

A perpendicular line has the negative reciprocal as its slope

$$\therefore m = \frac{-1}{-1.128} = +.887$$

Through the pt $\left(\frac{1}{2}, -\frac{1}{2}\right)$

$$y - y_1 = m(x - x_1)$$

$$\left(y + \frac{1}{2}\right) = .89\left(x - \frac{1}{2}\right)$$

47. **(C)** By substituting $y^2 = 4x$ into $x^2 + y^2 = 25$, we obtain $x^2 + 4x - 25 = 0$
Solve by the quadratic formula.

$$x = \frac{-4 \pm \sqrt{116}}{2} = -2 \pm \sqrt{29}$$

One root is positive and the other negative. Since $y = \pm 2\sqrt{x}$, the negative value of x gives us imaginary values of y, but the positive value of x gives us two real values of y. Hence, there are *two* points of intersection.

or

$x^2 + y^2 = 25$ is a circle with center at the origin and a radius of 5
$y^2 = 4x$ is a parabola with vertex at (0,0) open to the right

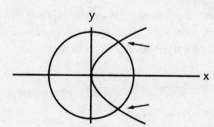

2 points of intersection

48. **(A)**

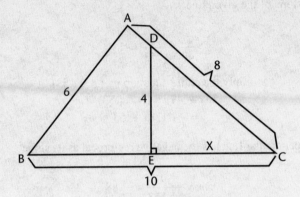

Since $\triangle ABC \sim \triangle DEC$, we may obtain the proportion

$$\frac{4}{6} = \frac{x}{8}$$

Cross-multiplying

$$6x = 32$$

$$x = 5\frac{1}{3}$$

49. **(D)** The man does $\frac{1}{h}$ of the job in 1 hour. The son does $\frac{1}{2h}$ of the job in 1 hour. Together they do $\frac{1}{h} + \frac{1}{2h} = \frac{3}{2h}$ of the job in 1 hour. Therefore, in $\frac{2h}{3}$ hours they will do $\frac{3}{2h} \cdot \frac{2h}{3} = 1$ complete job together. Therefore, it takes them $\frac{2h}{3}$ hours.

50. **(C)**

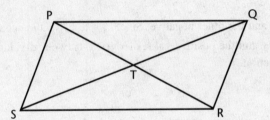

$\triangle PQT \cong \triangle STR$ and $\triangle PST \cong \triangle QRT$

In $\triangle PQS$, PT is a median, since the diagonals of a ▢ bisect each other. Consequently, $\triangle PST$ and $\triangle PTQ$ are equal in area as they have equal bases ($ST = TQ$) and a common altitude (perpendicular from P to SQ). Therefore, all four triangles are equal in area.

Sample Test 2
Answer Sheet

Math Level IC

1. Ⓐ Ⓑ Ⓒ Ⓓ Ⓔ
2. Ⓐ Ⓑ Ⓒ Ⓓ Ⓔ
3. Ⓐ Ⓑ Ⓒ Ⓓ Ⓔ
4. Ⓐ Ⓑ Ⓒ Ⓓ Ⓔ
5. Ⓐ Ⓑ Ⓒ Ⓓ Ⓔ
6. Ⓐ Ⓑ Ⓒ Ⓓ Ⓔ
7. Ⓐ Ⓑ Ⓒ Ⓓ Ⓔ
8. Ⓐ Ⓑ Ⓒ Ⓓ Ⓔ
9. Ⓐ Ⓑ Ⓒ Ⓓ Ⓔ
10. Ⓐ Ⓑ Ⓒ Ⓓ Ⓔ
11. Ⓐ Ⓑ Ⓒ Ⓓ Ⓔ
12. Ⓐ Ⓑ Ⓒ Ⓓ Ⓔ
13. Ⓐ Ⓑ Ⓒ Ⓓ Ⓔ
14. Ⓐ Ⓑ Ⓒ Ⓓ Ⓔ
15. Ⓐ Ⓑ Ⓒ Ⓓ Ⓔ
16. Ⓐ Ⓑ Ⓒ Ⓓ Ⓔ
17. Ⓐ Ⓑ Ⓒ Ⓓ Ⓔ
18. Ⓐ Ⓑ Ⓒ Ⓓ Ⓔ
19. Ⓐ Ⓑ Ⓒ Ⓓ Ⓔ
20. Ⓐ Ⓑ Ⓒ Ⓓ Ⓔ
21. Ⓐ Ⓑ Ⓒ Ⓓ Ⓔ
22. Ⓐ Ⓑ Ⓒ Ⓓ Ⓔ
23. Ⓐ Ⓑ Ⓒ Ⓓ Ⓔ
24. Ⓐ Ⓑ Ⓒ Ⓓ Ⓔ
25. Ⓐ Ⓑ Ⓒ Ⓓ Ⓔ
26. Ⓐ Ⓑ Ⓒ Ⓓ Ⓔ
27. Ⓐ Ⓑ Ⓒ Ⓓ Ⓔ
28. Ⓐ Ⓑ Ⓒ Ⓓ Ⓔ
29. Ⓐ Ⓑ Ⓒ Ⓓ Ⓔ
30. Ⓐ Ⓑ Ⓒ Ⓓ Ⓔ
31. Ⓐ Ⓑ Ⓒ Ⓓ Ⓔ
32. Ⓐ Ⓑ Ⓒ Ⓓ Ⓔ
33. Ⓐ Ⓑ Ⓒ Ⓓ Ⓔ
34. Ⓐ Ⓑ Ⓒ Ⓓ Ⓔ
35. Ⓐ Ⓑ Ⓒ Ⓓ Ⓔ
36. Ⓐ Ⓑ Ⓒ Ⓓ Ⓔ
37. Ⓐ Ⓑ Ⓒ Ⓓ Ⓔ
38. Ⓐ Ⓑ Ⓒ Ⓓ Ⓔ
39. Ⓐ Ⓑ Ⓒ Ⓓ Ⓔ
40. Ⓐ Ⓑ Ⓒ Ⓓ Ⓔ
41. Ⓐ Ⓑ Ⓒ Ⓓ Ⓔ
42. Ⓐ Ⓑ Ⓒ Ⓓ Ⓔ
43. Ⓐ Ⓑ Ⓒ Ⓓ Ⓔ
44. Ⓐ Ⓑ Ⓒ Ⓓ Ⓔ
45. Ⓐ Ⓑ Ⓒ Ⓓ Ⓔ
46. Ⓐ Ⓑ Ⓒ Ⓓ Ⓔ
47. Ⓐ Ⓑ Ⓒ Ⓓ Ⓔ
48. Ⓐ Ⓑ Ⓒ Ⓓ Ⓔ
49. Ⓐ Ⓑ Ⓒ Ⓓ Ⓔ
50. Ⓐ Ⓑ Ⓒ Ⓓ Ⓔ

Directions: For each question in the sample test, select the best of the answer choices and blacken the corresponding space on this answer sheet.

Please note: (a) You will need to use a calculator in order to answer some, though not all, of the questions in this test. As you look at each question, you must decide whether or not you need a calculator for the specific question. A four-function calculator is not sufficient; your calculator must be at least a scientific calculator. Calculators that can display graphs and programmable calculators are also permitted.

(b) The only angle measure used on the Level IC test is degree measure. Your calculator should be set to degree mode.

(c) All figures are accurately drawn and are intended to supply useful information for solving the problems that they accompany. Figures are drawn to scale UNLESS it is specifically stated that a figure is not drawn to scale. Unless otherwise indicated, all figures lie in a plane.

(d) The domain of any function f is assumed to be the set of all real numbers x for which $f(x)$ is a real number except when this is specified not to be the case.

(e) Use the reference data below as needed.

REFERENCE DATA

Solid	Volume	Other		
Right circular cone	$V = \frac{1}{3}\pi r^2 h$	$S = cl$	V = volume r = radius h = height	S = lateral area c = circumference of base l = slant height
Sphere	$V = \frac{4}{3}\pi r^3$	$S = 4\pi r^2$	V = volume r = radius S = surface area	
Pyramid	$V = \frac{1}{3}Bh$		V = volume B = area of base h = height	

Sample Test 2

MATH LEVEL IC

50 Questions • Time—60 Minutes

DO YOUR FIGURING HERE.

1. $\sqrt{13^2 - 12^2} = \sqrt[n]{125}$, $n =$

    (A) 1   (B) 2   (C) 3   (D) 4   (E) 5

2. If $f(t) = 7t + 12$, for what value(s) of t is $f(t) > 33$?

    (A) $t > 3$ (B) $t < 3$ (C) $t = 3$ (D) $-3 < t < 3$
    (E) all values of t

3. If $(3, -5)$ are the coordinates of one endpoint of a diameter of a circle with center at $(6, 2)$, the coordinates of the other endpoint of this diameter are

    (A) $(9, -3)$ (B) $\left(4\frac{1}{2}, -1\frac{1}{2}\right)$ (C) $(3, 7)$ (D) $(9, 7)$ (E) $(9, 9)$

4. What is the slope of a line perpendicular to the line $\sqrt{11}x + \sqrt{5}y = 2$?

    (A) .67
    (B) .93
    (C) 1.07
    (D) 1.53
    (E) 1.82

5. If $f(x) = 3x - 5$ and $g(x) = x^2 + 1$, $f[g(x)] =$

    (A) $3x^2 - 5$ (B) $3x^2 + 6$ (C) $x^2 - 5$ (D) $3x^2 - 2$
    (E) $3x^2 + 5x - 2$

GO ON TO THE NEXT PAGE

6. In the right △ in figure 6, $60 > r > 45$. Which is true of x?

   (A) $10 < x < 10\sqrt{2}$
   (B) $5 < x < 5\sqrt{3}$
   (C) $10 > x > 5\sqrt{2}$
   (D) $10 > x > 5$
   (E) $5 < x < 5\sqrt{2}$

DO YOUR FIGURING HERE.

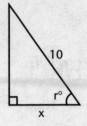

Fig. 6

7. Two planes are perpendicular to each other. The locus of points at distance d from one of these planes and at distance e from the other is

   (A) 2 lines (B) 4 lines (C) 2 points (D) 4 points (E) 2 planes

8. The graph of the equation $y = -2x + 3$ lies in quadrants

   (A) I and II only
   (B) I and III only
   (C) II and III only
   (D) I, II and IV only
   (E) II, III and IV only

9. A circle of radius .39 is circumscribed about a square. What is the perimeter of the square?

   (A) .78
   (B) 1.21
   (C) 1.78
   (D) 2.21
   (E) 2.78

10. The set of odd integers is closed under

    (A) addition (B) subtraction (C) multiplication (D) division
    (E) none of these

11. If $|2x + 3| \leq 9$ and $2x + 3 < 0$, then

    (A) $x \geq -6$ (B) $x < -\dfrac{3}{2}$ (C) $x \leq 3$ (D) $-6 \leq x \leq 3$
    (E) $-6 \leq x < -\dfrac{3}{2}$

12. The reciprocal of $6-2\sqrt{5}$ is equal to

    (A) $\dfrac{3-\sqrt{5}}{8}$ (B) $\dfrac{3+\sqrt{5}}{8}$ (C) $\dfrac{6+2\sqrt{5}}{41}$ (D) $\dfrac{6-2\sqrt{5}}{56}$

    (E) $\dfrac{6+2\sqrt{5}}{56}$

13. If the equation $9x^2 - 4Kx + 4 = 0$ has two equal roots, K is equal to

    (A) ±1   (B) 2   (C) ±3   (D) ±4   (E) 5

14. How many points of intersection are between the graphs of the equations $x^2 + y^2 = 7$ and $x^2 - y^2 = 1$?

    (A) 0   (B) 1   (C) 2   (D) 3   (E) 4

15. The sum of all numbers of the form $2K + 1$, where K takes on integral values from 1 to n is

    (A) n^2 (B) $n(n + 1)$ (C) $n(n + 2)$ (D) $(n + 1)^2$
    (E) $(n + 1)(n + 2)$

16. Which of the following properties of zero is the basis for excluding "division by zero"?

    (A) $K + 0 = K$ for every integer K
    (B) $K + (-K) = 0$ for every integer K
    (C) 0 is its own additive inverse
    (D) $K \cdot 0 = 0$ for every integer K
    (E) 0 is the additive identity

17. The graph of the hyperbola $\dfrac{x^2}{25} - \dfrac{y^2}{9} = 1$ has no points in the vertical strip between the lines

    (A) $x = 9$ and $x = -9$
    (B) $x = 3$ and $x = -3$
    (C) $x = 25$ and $x = -25$
    (D) $x = 5$ and $x = -5$
    (E) $x = 10$ and $x = -10$

18. If $P = EI$ and $E = IR$, $P =$

    (1) I^2R (2) $\dfrac{I}{R}$ (3) E^2R (4) $\dfrac{E^2}{R}$

    (A) 1 only   (B) 4 only   (C) 1 and 4 only   (D) 1 and 3 only
    (E) 2 and 4 only

DO YOUR FIGURING HERE.

19. In △PQR, M is the midpoint of PQ, and N is the midpoint of QR. MS ⊥ PR and NT ⊥ PR. The sum of the areas of △MPS and △NTR is what part of the area of △PQR?

   (A) $\frac{1}{2}$ (B) $\frac{1}{3}$ (C) $\frac{1}{4}$ (D) $\frac{2}{5}$ (E) $\frac{3}{8}$

DO YOUR FIGURING HERE.

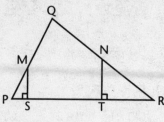

Fig. 19

20. A man usually gets from one corner of a square lot to the opposite corner by walking along two of the sides. Approximately what percent of the distance does he save if he walks along the diagonal?

   (A) 27%   (B) 29%   (C) 31%   (D) 33%   (E) 25%

21. In figure 21, all intersections occur at right angles. What is the perimeter of the polygon?

   (A) 20.99
   (B) 20.19
   (C) 19.38
   (D) 19.27
   (E) 15.11

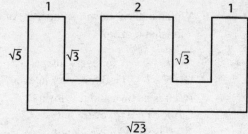

Fig. 21

22. Solve the equation $p = \frac{ry}{r+y}$ for y.

   (A) $y = \frac{r+p}{rp}$ (B) $y = \frac{rp}{r+p}$ (C) $y = \frac{r-p}{rp}$ (D) $y = \frac{rp}{r-p}$

   (E) $y = \frac{r+p}{r-p}$

23. If 6.12% of $\frac{7}{13}$ is .43 percent of x, what is the value of x?

   (A) .077
   (B) .77
   (C) 7.66
   (D) 8.13
   (E) 11.49

24. If x is a positive acute angle and $\tan x = p$, $\cos x$ is equal to

    (A) $\dfrac{p}{\sqrt{p^2+1}}$ (B) $\sqrt{p^2+1}$ (C) $\dfrac{1}{\sqrt{p^2+1}}$ (D) $\dfrac{1}{\sqrt{p^2-1}}$

    (E) $\dfrac{p}{\sqrt{p^2-1}}$

25. If p lb of salt are dissolved in q lb of water, the percent of salt in the resulting solution is

    (A) $\dfrac{100p}{p+q}$ (B) $\dfrac{p}{p+q}$ (C) $\dfrac{100p}{q}$ (D) $\dfrac{100q}{p+q}$
    (E) none of these

26. How many numbers between 131 and 259 are divisible by 3?

    (A) 41 (B) 42 (C) 43 (D) 44 (E) 45

27. If $f(x) = x^2 - 3x + 4$, which of the following is equal to $f(f(f(3\tfrac{7}{8})))$?

    (A) −65.32
    (B) −18.41
    (C) 86.89
    (D) 1123.78
    (E) 1223.21

28. The fraction $\dfrac{m^{-1}}{m^{-1}+n^{-1}}$ is equal to

    (A) m (B) $\dfrac{m}{m+n}$ (C) $\dfrac{m+n}{m}$ (D) $\dfrac{n}{m+n}$
    (E) $\dfrac{m+n}{m-n}$

29. The graph of the equation $4y^2 + x^2 = 25$ is

    (A) a circle (B) an ellipse (C) a hyperbola (D) a parabola
    (E) a straight line

30. If $f(x) = \dfrac{x+1}{x-1}$ and $g(x) = 2x - 1$, $f[g(x)] =$

    (A) $\dfrac{x-1}{x}$ (B) $\dfrac{x}{x+1}$ (C) $\dfrac{x+1}{x}$ (D) $\dfrac{x}{x-1}$ (E) $\dfrac{2x-1}{2x+1}$

31. A circle of radius x has an area twice that of a square of side a. The equation used to find the radius of the circle is

    (A) $\pi a^2 = 2x^2$
    (B) $\pi x^2 = 2a^2$
    (C) $\pi x^2 = 4a^2$
    (D) $\pi a^2 = 4x^2$
    (E) $4\pi x^2 = a^2$

32. In figure 32, the shaded region within the triangle is the intersection of the sets of ordered pairs described by which of the following inequalities?

    (A) $y < x, x < 2$
    (B) $y < 2x, x < 2$
    (C) $y < 2x, x < 2, x > 0$
    (D) $y < 2x, y < 2, x > 0$
    (E) $y < 2x, x < 2, y > 0$

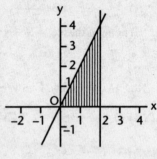

Fig. 32

33. How many digits are in the number $(63)^{21}$?

    (A) 21
    (B) 37
    (C) 38
    (D) 39
    (E) 63

34. If the roots of the equation $2x^2 - 3x + c = 0$ are real and irrational, a possible value of c is

    (A) -2 (B) -1 (C) 0   (D) 1   (E) 2

35. If the graphs of $x - ay = 10$ and $2x - y = 3$ are perpendicular lines, $a =$

    (A) 0   (B) -1 (C) -2 (D) 1   (E) 2

36. If a chord 8 inches long has an arc of 120°, the radius, in inches, of the circle is

    (A) 4   (B) $4\sqrt{3}$ (C) $\frac{8\sqrt{3}}{5}$ (D) $\frac{8\sqrt{3}}{3}$ (E) $2\sqrt{3}$

37. If $4m = 5K$ and $6n = 7K$, the ratio of m to n is

    (A) 5:7   (B) 10:21   (C) 14:15   (D) 2:3   (E) 15:14

38. The formula $F = \frac{9}{5}C + 32$ relates any Centigrade reading to its corresponding Fahrenheit reading. For what temperature is the reading the same on both scales?

    (A) 40°   (B) 0°   (C) −40°   (D) −32°   (E) −73°

39. Which of the following is a root of $y = 3x^2 - 17x + 5$?

    (A) −.31
    (B) .41
    (C) 2.11
    (D) 3.56
    (E) 5.36

40.

x	−1	0	3	6
y	−1	1	7	13

A linear equation expressing the relation between x and y in the above table is

(A) $y = 2x - 1$ (B) $y = 2x + 1$ (C) $y = 2x$ (D) $2x + 3y = 5$
(E) $2x - y = 4$

41. Find the smallest number (other than 2) which, when divided by 3, 4, 5, 6, or 7, leaves a remainder of 2.

    (A) 422   (B) 842   (C) 2002   (D) 2522   (E) 5102

42. $\dfrac{8^5 \cdot 9^4}{2^{12} \cdot 3^6} =$

    (A) 48   (B) 54   (C) 1152   (D) 128   (E) 72

DO YOUR FIGURING HERE.

43. In a class of 250 students, 175 take mathematics and 142 take science. How many take both mathematics and science? (All take math and/or science.)

    (A) 67
    (B) 75
    (C) 33
    (D) 184
    (E) cannot be determined from information given

44. The coordinates of vertices P and Q of an equilateral triangle PQR are $(-5, 2)$ and $(5, 2)$ respectively. The coordinates of R may be

    (A) $(0, 5\sqrt{3})$ (B) $(0, -5\sqrt{3})$ (C) $(0, 2+5\sqrt{3})$ (D) $(0, 7\sqrt{3})$
    (E) $(-2, 5\sqrt{3})$

45. If $r - s > r + s$, then

    (A) $r > s$ (B) $s < 0$ (C) $r < 0$ (D) $r > s$ (E) $s > 0$

46. In figure 46, for all triangles ABC

    (A) $\angle x > \angle A$
    (B) $\angle x < \angle A$
    (C) $\angle x > \angle y$
    (D) $\angle x < \angle y$
    (E) $\angle D > \angle A$

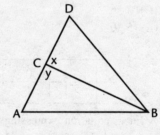

Fig. 46

47. Given: All seniors are mature students. Which statement expresses a conclusion that logically follows from the given statement?

    (A) All mature students are seniors.
    (B) If Bill is a mature student, then he is a senior.
    (C) If Bill is not a mature student, then he is not a senior.
    (D) If Bill is not a senior, then he is not a mature student.
    (E) All sophomores are not mature students.

48. In figure 48, AB is parallel to CD. How many degrees in angle EFG?

    (A) 90°
    (B) 10°
    (C) 60°
    (D) 70°
    (E) cannot be determined from information given

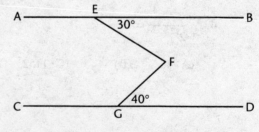

Fig. 48

49. The volume V of a circular cylinder of radius r and altitude h is given by the formula $V = \pi r^2 h$. Which, if any, of the following statements is true?

   (A) If both r and h are doubled, V is doubled.
   (B) If r is increased by 2, V is increased by 4.
   (C) If r is doubled and h is halved, V remains the same.
   (D) If r is doubled and h is divided by 4, V remains the same.
   (E) none of these

50. In $\triangle PQR$, the sides are 10, 17, and 21 and $PS \perp QR$. The length of PS is

   (A) $6\sqrt{3}$
   (B) $6\frac{2}{5}$
   (C) $6\sqrt{2}$
   (D) $\sqrt{63}$
   (E) 8

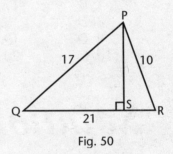

Fig. 50

STOP

Sample Test 2
Answer Key

Math Level IC

1. C	11. E	21. A	31. B	41. A
2. A	12. B	22. D	32. E	42. E
3. E	13. C	23. C	33. C	43. A
4. A	14. E	24. C	34. B	44. C
5. D	15. C	25. A	35. C	45. B
6. E	16. D	26. C	36. D	46. A
7. B	17. D	27. E	37. E	47. C
8. D	18. C	28. D	38. C	48. D
9. D	19. C	29. B	39. E	49. D
10. C	20. B	30. D	40. B	50. E

Solutions

1. **(C)** $\sqrt{13^2 - 12^2} = \sqrt{169 - 144} = \sqrt{25} = 5$

   $\sqrt[n]{125} = (125)^{1/n} = (5^3)^{1/n} = 5^{3/n} = 5^1$

   Since the bases are equal, set the exponents equal.

   $$\frac{3}{n} = 1$$
   $$n = 3$$

2. **(A)** $f(t) = 7t + 12 > 33$

   Subtract 12 from both sides.

   $$7t > 21$$
   $$t > 3$$

3. **(E)** Call coordinates of other endpoint (x, y).

   $6 = \dfrac{3+x}{2}$ $2 = \dfrac{-5+y}{2}$

   $12 = 3 + x$ $4 = -5 + y$

   $x = 9$ $y = 9$

   Other endpoint is (9, 9).

164

4. **(A)**
$$\sqrt{11}x + \sqrt{5}y = 2$$
$$\Downarrow$$
$$\sqrt{5}y = -\sqrt{11}x + 2$$
$$y = \frac{-\sqrt{11}}{\sqrt{5}} + \frac{2}{\sqrt{5}}$$
$$\text{slope} = m = \frac{-\sqrt{11}}{\sqrt{5}}$$

The slope of a line perpendicular to $m = \frac{-\sqrt{11}}{\sqrt{5}}$ has a slope of $\frac{+\sqrt{5}}{\sqrt{11}} = .674$. This is the negative reciprocal of $\frac{-\sqrt{11}}{\sqrt{5}}$.

5. **(D)** $f(x) = 3x - 5$, $g(x) = x^2 + 1$
$$f[g(x)] = 3[g(x)] - 5$$
$$= 3(x^2 + 1) - 5$$
$$= 3x^2 + 3 - 5$$
$$= 3x^2 - 2$$

6. **(E)**

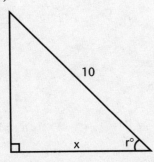

When $r = 60°$,
$$x = \frac{1}{2}(10) = 5.$$

When $r = 45°$,
$$x = \frac{1}{2}(10)\sqrt{2} = 5\sqrt{2}$$

If $60 > r > 45$,
$$5 < x < 5\sqrt{2}$$

7. **(B)** The locus of points at distance d from one plane consists of two planes parallel to the given plane and distance d from it.

The locus of points at distance e from the second plane consists of two planes parallel to the second plane and at distance e from it.

Since the two original planes are perpendicular to each other, the first pair of planes is perpendicular to the second pair of planes, resulting in 4 lines of intersection.

8. **(D)** 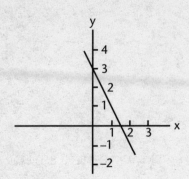 The graph of $y = -2x + 3$ has a slope of -2 and a y-intercept of 3. By setting $y = 0$, we see that the x-intercept is $1\frac{1}{2}$. Thus, the graph lies in quadrants I, II, and IV only.

9. **(D)**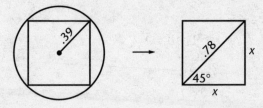

$$x^2 + x^2 = (.78)^2$$

$$x = \frac{.78}{\sqrt{2}}$$

Perimeter of square = $4x$

$$= 4\left(\frac{.78}{\sqrt{2}}\right) = 2.206$$

10. **(C)** Let x and y be two *odd* integers and represent $x = 2a + 1$ and $y = 2b + 1$. Where a and b are *any* integers,

$$\begin{aligned} x + y &= 2a + 1 + 2b + 1 \\ &= 2(a + b + 1) = \text{an even number} \\ x - y &= (2a + 1) - (2b + 1) \\ &= 2(a - b) = \text{an even number} \\ x \cdot y &= (2a + 1)(2b + 1) = 4ab + 2a + 2b + 1 \\ &= 2(2ab + a + b) + 1 \\ &= \text{an even number} + 1 = \text{an odd number} \end{aligned}$$

$\frac{x}{y}$ need not be an integer at all.

There is closure of the odd integers only under *multiplication*.

11. **(E)** If $2x + 3 < 0$, $|2x + 3| = -(2x + 3)$

$$|2x + 3| \leq 9$$
$$= -2x - 3 \leq 9$$
$$= -2x \leq 12$$
$$= x \geq -6$$

If $2x + 3 < 0$, $2x < -3$ and $x < -\frac{3}{2}$

Therefore the solution set is represented by

$$-6 \leq x < -\frac{3}{2}$$

12. **(B)** Rationalize the denominator of $\frac{1}{6 - 2\sqrt{5}}$.

$$\frac{1}{6 - 2\sqrt{5}} \cdot \frac{6 + 2\sqrt{5}}{6 + 2\sqrt{5}} = \frac{6 + 2\sqrt{5}}{6^2 - (2\sqrt{5})^2}$$

$$= \frac{6 + 2\sqrt{5}}{36 - 20} = \frac{6 + 2\sqrt{5}}{16}$$

$$= \frac{2(3 + \sqrt{5})}{16} = \frac{3 + \sqrt{5}}{8}$$

13. **(C)** If the roots of $9x^2 - 4Kx + 4 = 0$ are equal, the discriminant must be zero.

$$16K^2 - 144 = 0$$
$$16K^2 = 144$$
$$K^2 = 9$$
$$K = \pm 3$$

14. **(E)** Make a sketch of the graphs of the two equations.

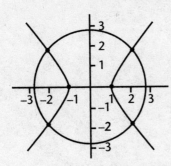

The graph of $x^2 + y^2 = 7$ is a circle with center at $(0, 0)$ and radius = $\sqrt{7} \approx 2.6$.

The graph of $x^2 - y^2 = 1$ is a hyperbola with intercepts at $(1, 0)$ and $(-1, 0)$. Thus there are 4 points of intersection.

15. **(C)** As *K* varies from 1 to *n*, $(2K+1)$ becomes $3, 5, 7, 9 \ldots (2n+1)$.
    This is an arithmetic progression with first term $a = 3$, last term $l = 2n + 1$.
    The sum *S* is given by the formula:

    $$S = \frac{n}{2}(a + 1)$$

    $$S = \frac{n}{2}(3 + 2n + 1)$$

    $$S = \frac{n}{2}(2n + 4)$$

    $$S = n(n + 2)$$

16. **(D)** If we tried to divide a number *n* by 0, and defined the quotient as a number *K*, $\frac{n}{0} = K$ and $k \cdot 0$ equals *n*. But $K \cdot 0$ is always 0, and we could not get any number *n* as the product.

17. **(D)** Solve the equation $\frac{x^2}{25} - \frac{y^2}{9} = 1$ for *y*.

    Multiply by 225.

    $$9x^2 - 25y^2 = 225$$

    $$25y^2 = 9x^2 - 225$$

    $$y^2 = \frac{9x^2 - 225}{25}$$

    $$y = \pm \frac{1}{5}\sqrt{9x^2 - 225}$$

    y is imaginary if $9x^2 - 225 < 0$
    $$9x^2 < 225$$
    $$x^2 < 25$$
    $$|x| < 5$$

    Thus there are no points in the vertical strip between $x = 5$ and $x = -5$.

18. **(C)** $P = EI$, $E = IR$
    Substitute $E = IR$ in first equation.
    $$P = IR \cdot I = I^2 R \quad (1)$$

    Also, solve second equation for *I* and substitute in first equation $I = \frac{E}{R}$.

    $$P = E \cdot \frac{E}{R} = \frac{E^2}{R} \quad (4)$$

    (1) and (4) only are possible.

19. **(C)** Draw $QV \perp PR$. The area of $\triangle PQR = \frac{1}{2} PR \cdot QV$

Since $MS \parallel QV \parallel NT$ and M and N are midpoints, it follows that $MS = NT = \frac{1}{2} QV$

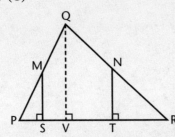

Also, $PS = \frac{1}{2} PV$ and $TR = \frac{1}{2} VR$

Area of $\triangle NTR = \frac{1}{2} NT \cdot RT$

$= \frac{1}{2} \left(\frac{1}{2} QV \right) \left(\frac{1}{2} PV \right) = \frac{1}{4}$ area of $\triangle PQV$

and Area of $\triangle NTR = \frac{1}{2} NT \cdot RT$

$= \frac{1}{2} \left(\frac{1}{2} QV \right) \left(\frac{1}{2} VR \right) = \frac{1}{4}$ area of $\triangle RQV$

Add these two equalities.

Area of $\triangle MPS + \triangle NTR = \frac{1}{4}$ area of $(\triangle PQV + \triangle RQV)$

$= \frac{1}{4}$ area of $\triangle PQR$

20. **(B)** $PQ + QR = 2, PR = \sqrt{2}$

Distance saved $= 2 - \sqrt{2}$

Percent saved $= \frac{2 - \sqrt{2}}{2} \times 100\%$

$= \frac{2 - 1.414}{2} \times 100\%$

$= \frac{.586}{2} \times 100\%$

$= .293 \times 100\%$

$= 29.3\% \approx 29\%$

21. **(A)**

The ___ $= \sqrt{23}$

Perimeter $= \sqrt{23} + \sqrt{5} + 4\sqrt{3} + \sqrt{23} + \sqrt{5}$

$= 2\sqrt{23} + 2\sqrt{5} + 4\sqrt{3}$

$= 20.992$

22. (D)

$$p = \frac{ry}{r+y}$$

Multiply both sides by $r + y$.

$$p(r + y) = ry$$
$$pr + py = ry$$
$$pr = ry - py.$$

Factor the right hand member.

$$pr = y(r - p)$$

Divide both sides by $(r - p)$.

$$y = \frac{pr}{r - p}$$

23. (C)

$$\frac{6.12}{100}\left(\frac{7}{13}\right) = \frac{.43}{100}(x)$$

$$.03295 = .0043x$$

$$x = 7.66$$

24. (C)

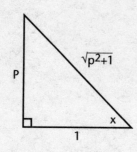

Construct a right $\triangle$ with one acute angle $= x$, and designate the opposite leg as p and the adjacent leg as 1, so that

$$\tan x = \frac{p}{1} = p.$$

By the Pythagorean Theorem, the hypotenuse is equal to $\sqrt{p^2 + 1}$ and $\cos x = \dfrac{1}{\sqrt{p^2 + 1}}$

25. (A) Total amount of solution $= (p + q)$ lb

$$\text{Percent of salt} = \frac{p}{p + q} \times 100$$

$$= \frac{100p}{p + q}$$

26. (C) 131, 132, 135 ... 258, 259
Divide by 3.
44, 45 ... 86

There are as many numbers divisible by 3 as are numbers between 44 and 86 inclusive.

$$86 - 44 + 1 = 43$$

27. **(E)**
$$f(x) = x^2 - 3x + 4$$
$$f\left(3\frac{7}{8}\right) = f(3.3875) = 7.39$$
$$f(7.39) = 36.45$$
$$f(36.45) = 1223.2$$

28. **(D)** $\dfrac{m^{-1}}{m^{-1}+n^{-1}} = \dfrac{\frac{1}{m}}{\frac{1}{m}+\frac{1}{n}}$

Multiply by mn.

$$\dfrac{\frac{1}{m} \cdot mn}{mn\left(\frac{1}{m}+\frac{1}{n}\right)} = \dfrac{n}{n+m}$$

29. **(B)** The equation $x^2 + 4y^2 = 25$ is of the form

$$b^2x^2 + a^2y^2 = a^2b^2 \quad \text{or} \quad \dfrac{x^2}{a^2} + \dfrac{y^2}{b^2} = 1$$

which are the standard forms of an ellipse with center at the origin. The graph of the equation is therefore an ellipse.

30. **(D)**
$$f(x) = \dfrac{x+1}{x-1}, \ g(x) = 2x - 1$$

$$f[g(x)] = \dfrac{g(x)+1}{g(x)-1} = \dfrac{2x-1+1}{2x-1-1} = \dfrac{2x}{2x-2}$$

$$= \dfrac{x}{x-1}$$

31. **(B)** Let the radius of the circle $= x$. Then, the area of the circle $= \pi x^2$.
Since the area of the square is a^2, the equation expressing the given conditions of the problem is

$$\pi x^2 = 2a^2$$

32. **(E)** The set of points to the left of the line $x = 2$ is designated by the inequality $x < 2$. The set of points above the x-axis is designated by the inequality $y > 0$.

The set of points under the line $y = 2x$ is designated by the inequality $y < 2x$.

Hence, the sketched region is designated by the intersection of the sets described by the inequalities

$$y < 2x, \quad x < 2, y > 0$$

33. **(C)** $(63)^{21} = 6.11155 \times 10^{37}$

The decimal point is moved 37 places to the right.
This yields a 38-digit number.

34. **(B)** For the roots of $2x^2 - 3x + c = 0$ to be real and irrational, the discriminant must be positive but not a perfect square.

Hence, $9 - 8C > 0$ or $9 > 8C$ or $C < 1\frac{1}{8}$.

This applies to all listed values except 2. However, $C = -2$ makes the discriminant 25; $C = -1$ makes it 17; $C = 0$ makes it 9; and $C = 1$ makes it 1. All are perfect squares except 17. Hence, the only possible value of C is -1.

35. **(C)** Slopes are negative reciprocals.

$$x - ay = 10 \qquad\qquad 2x - y = 3$$
$$x - 10 = ay \qquad\qquad 2x - 3 = y$$
$$y = \frac{1}{a}x - \frac{10}{a}$$

slope is $\frac{1}{a}$ 　　　　　slope is 2

　　　　　　　　　　　thus $-a = 2$
　　　　　　　　　　　or $a = -2$

36. **(D)** 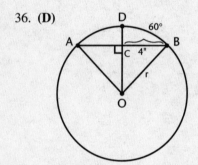 $OB = r$.

Draw $OC \perp AB$ and extend to point D on $\widehat{AB}$.
Then $DB = 60°$ and $\angle COB = 60°$
Thus $\angle B = 30°$.

In right triangle OBC, it follows that $BC = 4''$ and $OC = \frac{r}{2}$.

By the Pythagorean Theorem,

$$r^2 = \left(\frac{r}{2}\right)^2 + 4^2$$
$$r^2 = \frac{r^2}{4} + 16$$
$$4r^2 = r^2 + 64$$
$$3r^2 = 64$$
$$r^2 = \frac{64}{3}$$
$$r = \frac{8}{\sqrt{3}}$$
$$= \frac{8\sqrt{3}}{\sqrt{3}}$$

37. **(E)** If $4m = 5K$, $m = \frac{5}{4}K$

If $6n = 7K$, $n = \frac{7}{6}K$

$$\frac{m}{n} = \frac{\frac{5}{4}K}{\frac{7}{6}K}$$

Divide the numerator and denominator by K and simplify.

$$\frac{m}{n} = \frac{5}{4} \cdot \frac{6}{7} = \frac{15}{14}$$

38. **(C)** Let $F = C = x$, and substitute in

$$F = \frac{9}{5}C + 32$$

$$x = \frac{9}{5}x + 32$$

Multiply through by 5.

$$5x = 9x + 160$$
$$-4x = 160$$
$$x = -40$$

39. **(E)**
$$y = 3x^2 - 17x + 5$$

$$a = 3, \ b = -17, \ c = 5$$

$$x = \frac{-b \pm \sqrt{b^2 - 4ac}}{2a}$$

$$x = \frac{-(-17) \pm \sqrt{(-17)^2 - 4(3)(5)}}{2(3)}$$

$$x = \frac{17 \pm \sqrt{229}}{6}$$

$$x = \{.311, \ 5.36\}$$

40. **(B)** Since $y = 1$ when $x = 0$, the y-intercept is 1. We see from the table that as x increases 1 unit, y increases 2 units, so that the slope of the line must be two. It can be seen also from the table that the slope is constant, for when x increases 3 units, y increases 6 units, again yielding a slope of 2. The equation of the line =

$$y = 2x + 1$$

41. **(A)** Take the product of 3, 4, 5, and 7, which is divisible by 3, 4, 5, 6, and 7. Now add 2 to this product.

$$3 \cdot 4 \cdot 5 \cdot 7 + 2 = 420 + 2 = 422$$

42. **(E)**
$$\frac{8^2 \cdot 9^4}{2^{12} \cdot 3^6} = \frac{(2)^{3 \cdot 5} \cdot (3^2)^4}{2^{12} \cdot 3^6} = \frac{2^{15} \cdot 3^8}{2^{12} \cdot 3^6}$$

$$= \frac{2^{15}}{2^{12}} \cdot \frac{3^8}{3^6} = 2^3 \cdot 3^2$$

$$= 8 \cdot 9 = 72$$

43. **(A)**

MATH-175 SCIENCE-142 TOTAL NUMBER OF STUDENTS = 250

175 – x math only | x | 142 – x science only

x = No. of students taking both math and science.

$$250 = (175 - x) + x + (142 - x)$$
$$250 = 175 + (142 - x)$$
$$250 = 317 - x$$
$$x = 317 - 250 = 67$$

44. **(C)**

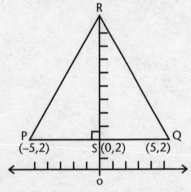

Since $\angle P$ is 60° and $PS = 5$, $PR = 10$ and $RS = 5\sqrt{3}$. Therefore, the coordinates of R are $(0, 2 + 5\sqrt{3})$.

45. **(B)** $r - s > r + s$
Add $-r$ to both sides.

$$-s > s$$

This can be true only if s is a negative number: $s < 0$.

46. **(A)** In the figure, $\angle x$ is an exterior angle of the triangle ABC and $\angle A$ is a non-adjacent interior angle. Therefore, for all triangles ABC

$$\angle x > \angle A$$

47. **(C)** (A) states a converse of the original statement and the converse need not follow
    (B) is essentially another form of the converse
    (C) is the contrapositive of the original and is equivalent to the original statement
    (D) is the inverse of the original statement and the inverse need not follow
    (E) is not implied by the given statement

48. **(D)**

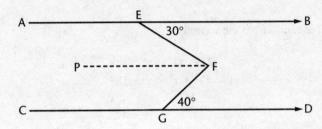

Draw $FP \parallel AB \parallel CD$.
By alternate interior angles of $\parallel$ lines,

$$\angle EFP = \angle BEF = 30°$$
and
$$\angle GFP = \angle FGD = 40°$$
$$\angle EFG = \angle EFP + \angle GFP$$
$$= 30° + 40°$$
$$= 70°$$

49. **(D)** If r is multiplied by 2, the effect is to multiply V by 4. If h is at the same time divided by 4, V remains the same.

50. **(E)** 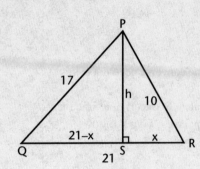 Let $RS = x$, $QS = 21 - x$, and $PS = h$.
By the Pythagorean Theorem
$h^2 + x^2 = 100$ and $h^2 + (21 - x)^2 = 289$

Subtract the first equation from the second.

$$(21 - x)^2 - x^2 = 189$$
$$441 - 42x + x^2 - x^2 = 189$$
$$441 - 189 = 42x$$
$$252 = 42x$$
$$x = 6$$

Thus,

$$h^2 = 100 - 36$$
$$h^2 = 64$$
$$h = 8$$

Sample Test 3
Answer Sheet

Math Level IC

1. Ⓐ Ⓑ Ⓒ Ⓓ Ⓔ
2. Ⓐ Ⓑ Ⓒ Ⓓ Ⓔ
3. Ⓐ Ⓑ Ⓒ Ⓓ Ⓔ
4. Ⓐ Ⓑ Ⓒ Ⓓ Ⓔ
5. Ⓐ Ⓑ Ⓒ Ⓓ Ⓔ
6. Ⓐ Ⓑ Ⓒ Ⓓ Ⓔ
7. Ⓐ Ⓑ Ⓒ Ⓓ Ⓔ
8. Ⓐ Ⓑ Ⓒ Ⓓ Ⓔ
9. Ⓐ Ⓑ Ⓒ Ⓓ Ⓔ
10. Ⓐ Ⓑ Ⓒ Ⓓ Ⓔ
11. Ⓐ Ⓑ Ⓒ Ⓓ Ⓔ
12. Ⓐ Ⓑ Ⓒ Ⓓ Ⓔ
13. Ⓐ Ⓑ Ⓒ Ⓓ Ⓔ
14. Ⓐ Ⓑ Ⓒ Ⓓ Ⓔ
15. Ⓐ Ⓑ Ⓒ Ⓓ Ⓔ
16. Ⓐ Ⓑ Ⓒ Ⓓ Ⓔ
17. Ⓐ Ⓑ Ⓒ Ⓓ Ⓔ
18. Ⓐ Ⓑ Ⓒ Ⓓ Ⓔ
19. Ⓐ Ⓑ Ⓒ Ⓓ Ⓔ
20. Ⓐ Ⓑ Ⓒ Ⓓ Ⓔ
21. Ⓐ Ⓑ Ⓒ Ⓓ Ⓔ
22. Ⓐ Ⓑ Ⓒ Ⓓ Ⓔ
23. Ⓐ Ⓑ Ⓒ Ⓓ Ⓔ
24. Ⓐ Ⓑ Ⓒ Ⓓ Ⓔ
25. Ⓐ Ⓑ Ⓒ Ⓓ Ⓔ
26. Ⓐ Ⓑ Ⓒ Ⓓ Ⓔ
27. Ⓐ Ⓑ Ⓒ Ⓓ Ⓔ
28. Ⓐ Ⓑ Ⓒ Ⓓ Ⓔ
29. Ⓐ Ⓑ Ⓒ Ⓓ Ⓔ
30. Ⓐ Ⓑ Ⓒ Ⓓ Ⓔ
31. Ⓐ Ⓑ Ⓒ Ⓓ Ⓔ
32. Ⓐ Ⓑ Ⓒ Ⓓ Ⓔ
33. Ⓐ Ⓑ Ⓒ Ⓓ Ⓔ
34. Ⓐ Ⓑ Ⓒ Ⓓ Ⓔ
35. Ⓐ Ⓑ Ⓒ Ⓓ Ⓔ
36. Ⓐ Ⓑ Ⓒ Ⓓ Ⓔ
37. Ⓐ Ⓑ Ⓒ Ⓓ Ⓔ
38. Ⓐ Ⓑ Ⓒ Ⓓ Ⓔ
39. Ⓐ Ⓑ Ⓒ Ⓓ Ⓔ
40. Ⓐ Ⓑ Ⓒ Ⓓ Ⓔ
41. Ⓐ Ⓑ Ⓒ Ⓓ Ⓔ
42. Ⓐ Ⓑ Ⓒ Ⓓ Ⓔ
43. Ⓐ Ⓑ Ⓒ Ⓓ Ⓔ
44. Ⓐ Ⓑ Ⓒ Ⓓ Ⓔ
45. Ⓐ Ⓑ Ⓒ Ⓓ Ⓔ
46. Ⓐ Ⓑ Ⓒ Ⓓ Ⓔ
47. Ⓐ Ⓑ Ⓒ Ⓓ Ⓔ
48. Ⓐ Ⓑ Ⓒ Ⓓ Ⓔ
49. Ⓐ Ⓑ Ⓒ Ⓓ Ⓔ
50. Ⓐ Ⓑ Ⓒ Ⓓ Ⓔ

Directions: For each question in the sample test, select the best of the answer choices and blacken the corresponding space on this answer sheet.

Please note: (a) You will need to use a calculator in order to answer some, though not all, of the questions in this test. As you look at each question, you must decide whether or not you need a calculator for the specific question. A four-function calculator is not sufficient; your calculator must be at least a scientific calculator. Calculators that can display graphs and programmable calculators are also permitted.

(b) The only angle measure used on the Level IC test is degree measure. Your calculator should be set to degree mode.

(c) All figures are accurately drawn and are intended to supply useful information for solving the problems that they accompany. Figures are drawn to scale UNLESS it is specifically stated that a figure is not drawn to scale. Unless otherwise indicated, all figures lie in a plane.

(d) The domain of any function f is assumed to be the set of all real numbers x for which $f(x)$ is a real number except when this is specified not to be the case.

(e) Use the reference data below as needed.

REFERENCE DATA

Solid	Volume	Other		
Right circular cone	$V = \frac{1}{3}\pi r^2 h$	$S = cl$	V = volume r = radius h = height	S = lateral area c = circumference of base l = slant height
Sphere	$V = \frac{4}{3}\pi r^3$	$S = 4\pi r^2$	V = volume r = radius S = surface area	
Pyramid	$V = \frac{1}{3}Bh$		V = volume B = area of base h = height.	

Sample Test 3

MATH LEVEL IC

50 Questions • Time—60 Minutes

1. If $.0000058 = 5.8 \times 10^n$, $n =$

    (A) -4 (B) -5 (C) -6 (D) -7 (E) 5

2. In triangle *NJL* the measure of angle *N* is 90° and the measure of angle *L* is 24°. If $\overline{NL} = 10$, what is the length of $\overline{JL}$?

    (A) 10.75
    (B) 10.85
    (C) 10.95
    (D) 11.05
    (E) 11.15

3. If $4x - 3 > x + 9$, then

    (A) $x > 2$ (B) $x > 3$ (C) $x > 4$ (D) $8 > x > 4$ (E) $x > 0$

4. $\left(\dfrac{1}{r} + \dfrac{1}{s}\right)\left(\dfrac{r}{r+s}\right) =$

    (A) $\dfrac{1}{r}$ (B) $\dfrac{r}{(r+s)^2}$ (C) $\dfrac{r}{s}$ (D) $\dfrac{s}{r}$ (E) $\dfrac{1}{s}$

DO YOUR FIGURING HERE.

GO ON TO THE NEXT PAGE

5. What is the value of $(\sin 17°)^2 + (\cos 17°)^2$?

   (A) .03
   (B) .18
   (C) .72
   (D) 1.00
   (E) 1.27

6. In figure 6, $\angle P = 2 \angle Q$ and $\angle QRS = 108°$. Triangle PQR is

   (A) isosceles
   (B) right
   (C) obtuse
   (D) scalene
   (E) equilateral

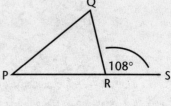

Fig. 6

7. For what value or values of y is the equation $\sqrt{y^2 + 27} = 2y$ satisfied?

   (A) ±3   (B) +3 only   (C) −3 only   (D) ±9   (E) +9 only

8. In figure 8, $\angle R > \angle T$ and RP and TP are bisectors of $\angle R$ and $\angle T$ respectively. Then

   (A) $PT < RP$
   (B) $PT = RP$
   (C) $RP + PT > RS + ST$
   (D) $PT > RP$
   (E) no relationship between PT and RP can be determined from the given information

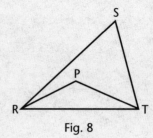

Fig. 8

9. If $x^5 - 8 = 159$, what is the value of x?

   (A) 2.67
   (B) 2.71
   (C) 2.78
   (D) 2.81
   (E) 2.84

10. In figure 10, $FG \parallel HK$, $FH \perp GH$, and $GK \perp HK$. If $FG = 5$ and $\angle F = r$, $HK =$

(A) $5 \sin^2 r$
(B) $5 \cos^2 r$
(C) $10 \sin^2 r$
(D) $5 \sin r$
(E) none of these

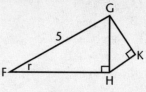

Fig. 10

11. If the graph of the equation $y = 2x^2 - 6x + C$ is tangent to the x-axis, the value of C is

(A) 3 (B) $3\frac{1}{2}$ (C) 4 (D) $4\frac{1}{2}$ (E) 5

12. If $\sqrt[3]{2x+4} = -.375$, then $x =$

(A) -2.03
(B) -1.97
(C) $-.87$
(D) $-.34$
(E) 1.43

13. In the formula $T = 2\pi \sqrt{\dfrac{L}{g}}$ π and g are constants. If we solve the formula for L,

(A) $\dfrac{Tg}{2\pi}$ (B) $\dfrac{Tg^2}{2\pi}$ (C) $\dfrac{T^2}{4\pi^2 g}$ (D) $\dfrac{T^2}{4\pi g^2}$ (E) $\dfrac{gT^2}{4\pi^2}$

14. A point P is 10 inches from a plane m. The locus of points in space which are 7 inches from P and 5 inches from plane m is

(A) a plane (B) a circle (C) two circles (D) a point
(E) two points

15. The equation of the graph in figure 15 is

    (A) $y = x + 1$
    (B) $y = |x - 1|$
    (C) $y = x^2 + 1$
    (D) $y = |x + 1|$
    (E) $y = |x|$

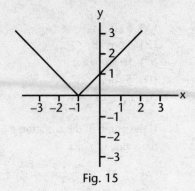

Fig. 15

16. Select the correct order for defining the following terms:

    I—natural number   II—imaginary number
    III—rational number   IV—integer

    (A) I, IV, III, II
    (B) I, II, III, IV
    (C) I, III, II, IV
    (D) IV, I, III, II
    (E) I, IV, II, III

17. If the reciprocal of $y - 1$ is $y + 1$, y equals

    (A) -1 (B) $+1$ (C) 0 (D) ± 1 (E) none of these

18. In a right triangle having angles of 30° and 60°, the 60° angle is bisected. What is the ratio of the segments into which the angle bisector divides the opposite leg?

    (A) 2:3   (B) 3:4   (C) 1:2   (D) 3:5   (E) 2:5

19. The equation $4y^2 - 3y + C = 0$ has real roots. The value of C for which the product of the roots is a maximum is

    (A) $\frac{9}{16}$ (B) $\frac{9}{4}$ (C) $\frac{4}{9}$ (D) $\frac{3}{4}$ (E) $-\frac{4}{3}$

20. The sum of all the *even* numbers between 1 and 51 is

    (A) 1300   (B) 650   (C) 325   (D) 675   (E) none of these

21. If $\frac{a}{b} = \frac{c}{d}$ (a, b, c, d positive numbers), which one of the following is not always true?

    (A) $\frac{a}{c} = \frac{b}{d}$ (B) $\frac{b}{a} = \frac{d}{c}$ (C) $\frac{a+b}{b} = \frac{c+d}{d}$ (D) $\frac{a}{d} = \frac{b}{c}$

    (E) $\frac{a}{b} = \frac{a+c}{b+d}$

22. The equation of the locus of points equidistant from $P(-2, -3)$ and $Q(-2, 5)$ is

(A) $y = 1$ (B) $y = -1$ (C) $x = 1$ (D) $x = -1$ (E) $y = -x$

23. If $f(x) = \frac{x+1}{x-1}$, what is the value of $f\left(f\left(f\left(f\left(\frac{3}{5}\right)\right)\right)\right)$?

(A) -4
(B) 0
(C) .6
(D) 1.3
(E) 7

24. The base of a triangle is 16 inches and its altitude is 10 inches. The area of the trapezoid cut off by a line 4 inches from the vertex is

(A) 134.4
(B) 67.2
(C) 38.6
(D) 72
(E) not determined from the information given

25. The locus of the centers of all circles of given radius r, in the same plane, passing through a fixed point P, is

(A) a straight line (B) two straight lines (C) a circle
(D) two circles (E) a point

26. The number of distinct points common to the graphs of $x^2 + y^2 = 4$ and $y^2 = 4$ is

(A) 0 (B) 1 (C) 2 (D) 3 (E) 4

27. Given the statement: All seniors are mature students. The statement that negates this statement is:

(A) All non-seniors are mature students.
(B) Some non-seniors are mature students.
(C) No seniors are mature students.
(D) All seniors are immature students.
(E) At least one senior is an immature student.

28. For what value of c is the parabola $y = 2.8x^2 - \sqrt{5}x + c$ tangent to the x-axis?

   (A) .35
   (B) .45
   (C) .55
   (D) .65
   (E) .75

29. The set of y-values that satisfies the inequality $|y - 5| < 6$ is

   (A) $1 < y < 11$ (B) $y > 11$ (C) $y < 11$ (D) $-1 < y < 11$
   (E) $|y| < 5$

30. The equation $r + \dfrac{5}{r-1} = 1 + \dfrac{5}{r-1}$ has

   (A) no root
   (B) one integral root
   (C) two equal roots
   (D) two unequal, rational roots
   (E) infinitely many roots

31. A cylindrical tank is $\dfrac{1}{2}$ full. When 6 quarts are added, the tank is $\dfrac{2}{3}$ full. The capacity of the tank, in quarts, is

   (A) 18   (B) 24   (C) 36   (D) 40   (E) 48

32. The diameters of two wheels are 10 in. and 14 in. The smaller makes 50 more revolutions than the larger in going a certain distance. This distance, in inches, is

   (A) 3500   (B) 1750   (C) 1750π (D) 3500π (E) none of these

33. The graphs of the equations $2x - 3y = 5$ and $4x - 6y = 7$

   (A) form an acute angle
   (B) intersect in two points
   (C) are parallel lines
   (D) are coincident lines
   (E) are perpendicular lines

DO YOUR FIGURING HERE.

34. In figure 34, what is the length of side $\overline{JL}$?

(A) 5.41
(B) 5.35
(C) 5.27
(D) 5.23
(E) 5.14

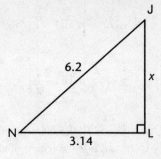

Fig. 34

35. Circles O and O' are tangent to each other. Circle O' passes through the center of O. If the area of circle O is 16, the area of circle O' is

(A) $2\sqrt{2}$
(B) 2
(C) $2\sqrt{\pi}$
(D) $4\sqrt{\pi}$
(E) 4

Fig. 35

36. In the right triangle in figure 36, $z = 26$ and $x - y = 14$.
$x + y =$

(A) 17
(B) 34
(C) 30
(D) 32
(E) 28

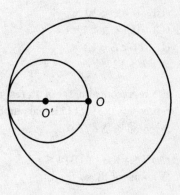

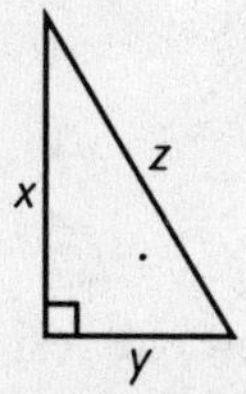

Fig. 36

37. A circle is inscribed in a square and then a smaller square is inscribed in the circle. The ratio of the area of the smaller square to that of the larger square is:

(A) 1:4 (B) $\sqrt{2}:2$ (C) 1:2 (D) $1:\sqrt{2}$ (E) 2:3

38. If $f(x) = x - \frac{1}{x}$, then $f\left(\frac{1}{x}\right) =$

I: $f(x)$ II: $f(-x)$ III: $-f(-x)$ IV: $\frac{1}{f(x)}$

(A) I and II (B) II and III (C) III and IV
(D) II only (E) II and IV

GO ON TO THE NEXT PAGE

39. Which one of the following is an irrational number?

   (A) $\sqrt[3]{-27}$ (B) $\sqrt{2}(3\sqrt{2}+2\sqrt{8})$ (C) $\dfrac{3\sqrt{18}}{2\sqrt{6}}$

   (D) $\sqrt{\dfrac{1}{2}} \cdot \sqrt{\dfrac{25}{2}}$ (E) $\dfrac{2\sqrt{5}}{\sqrt{45}}$

40. Quadrilateral $PQRS$ is inscribed in a circle of radius 10. If angle PQR is 150°, and L is the length of arc PQR, then

   (A) $L < 10$
   (B) $10 < L < 10.5$
   (C) $10.5 < L < 11$
   (D) $11 < L < 12$
   (E) $L > 12$

41. If S represents the set of all real numbers x such that $1 \leq x \leq 3$, and T represents the set of all real numbers x such that $2 \leq x \leq 5$, the set represented by $S \cap T$ is

   (A) $2 \leq x \leq 3$ (B) $1 \leq x \leq 5$ (C) $x \leq 5$ (D) $x \geq 1$
   (E) none of these

42. Which of the following is a zero of the equation $x^2 - 3x = 7$?

   (A) -4.54
   (B) -1.54
   (C) 1.54
   (D) 3.54
   (E) 5.54

43. A boy grew one year from a height of x inches to a height of y inches. The percent of increase was

   (A) $\dfrac{100(y-x)}{y}$ (B) $\dfrac{100(x-y)}{x}$ (C) $\dfrac{y-x}{x}$ (D) $\dfrac{100(y-x)}{x}$

   (E) $\dfrac{x-y}{x}$

44. In figure 44, how are the coordinates of P related?

   (A) $x < y$
   (B) $x > y$
   (C) $x = y$
   (D) $x \leq y$
   (E) $xy = 1$

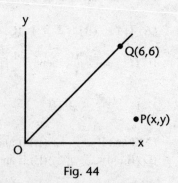

Fig. 44

45. A boy wishes to cut the largest possible square out of a piece of cardboard in the shape of a right triangle, with legs of 8 inches and 12 inches as shown in figure 45. The side of the square, in inches, is

(A) 4
(B) 5
(C) 4.8
(D) 4.5
(E) 4.3

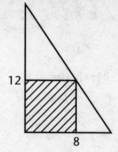

Fig. 45

Questions 46–50 pertain to the following situation: Two cubes have edges in the ratio of 2:3 respectively.

46. The ratio of their surface areas is

(A) $\frac{4}{9}$ (B) $\frac{8}{27}$ (C) $\frac{2}{3}$ (D) $\frac{\sqrt{2}}{\sqrt{3}}$ (E) $\frac{\sqrt{3}}{\sqrt{2}}$

47. The ratio of their volumes is

(A) $\frac{4}{9}$ (B) $\frac{8}{27}$ (C) $\frac{2}{3}$ (D) $\frac{\sqrt{2}}{\sqrt{3}}$ (E) $\frac{\sqrt{3}}{\sqrt{2}}$

48. The ratio of the sum of the edges of the smaller to the sum of the edges of the larger is

(A) $\frac{4}{9}$ (B) $\frac{8}{27}$ (C) $\frac{2}{3}$ (D) $\frac{\sqrt{2}}{\sqrt{3}}$ (E) $\frac{\sqrt{3}}{\sqrt{2}}$

49. The ratio of the diagonal of a face of the first cube to the diagonal of a face in the second is

(A) $\frac{4}{9}$ (B) $\frac{8}{27}$ (C) $\frac{2}{3}$ (D) $\frac{\sqrt{2}}{\sqrt{3}}$ (E) $\frac{\sqrt{3}}{\sqrt{2}}$

50. The ratio of a diagonal of the first cube to the diagonal of one of its faces is

(A) $\frac{4}{9}$ (B) $\frac{8}{27}$ (C) $\frac{2}{3}$ (D) $\frac{\sqrt{2}}{\sqrt{3}}$ (E) $\frac{\sqrt{3}}{\sqrt{2}}$

STOP

Sample Test 3
Answer Key

Math Level IC

1. C	11. D	21. D	31. C	41. A
2. C	12. A	22. A	32. C	42. B
3. C	13. E	23. C	33. C	43. D
4. E	14. B	24. B	34. B	44. B
5. D	15. D	25. C	35. E	45. C
6. A	16. A	26. C	36. B	46. A
7. B	17. E	27. E	37. C	47. B
8. D	18. C	28. B	38. B	48. C
9. C	19. A	29. D	39. C	49. C
10. A	20. B	30. A	40. B	50. E

Solutions

1. **(C)** $.0000058 = 5.8 \times 10^n = 5.8 \times 10^{-6}$, $n = -6$

2. **(C)**

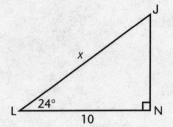

$$\cos 24° = \frac{10}{x}$$

$$x = \frac{10}{\cos 24°}$$

$$= 10.946$$

3. **(C)**
$$4x - 3 > x + 9$$
$$4x - x > 9 + 3$$
$$3x > 12$$
$$x > 4$$

4. **(E)**
$$\left(\frac{1}{r} + \frac{1}{s}\right)\left(\frac{r}{r+s}\right) = \frac{r+s}{rs} \cdot \frac{r}{r+s}$$

Dividing by r and $r + s$, we get $\frac{1}{s}$.

Solutions

5. **(D)**
$$(\sin 17°)^2 + (\cos 17°)^2$$
$$= \sin^2(17°) + \cos^2(17°)$$
$$= \sin^2\theta + \cos^2\theta$$
$$= 1$$

6. **(A)**
$$\angle QRS = \angle P + \angle Q$$
$$= 2\angle Q + \angle Q$$
$$108° = 3\angle Q$$
$$\angle Q = 36°$$
$$\angle P = 72°$$
$$\angle R = 180° - \angle QRS = 180° - 108° = 72°$$
$$\angle P = \angle R \text{ and } \triangle PQR \text{ is isosceles}$$

7. **(B)**
$$\sqrt{y^2 + 27} = 2y$$
$$y^2 + 27 = 4y^2$$
$$27 = 3y^2$$
$$y^2 = 9$$
$$y = \pm 3$$

Check $y = 3$
$\sqrt{9 + 27} = 6, \sqrt{36} = 6$ which checks.
Check $y = -3$.
$\sqrt{9 + 27} = -6, \sqrt{36} = -6$ which does not check.
Hence only $+3$ is a solution.

8. **(D)** Since $\angle R > \angle T$, $\frac{1}{2}\angle R > \frac{1}{2}\angle T$ or $\angle PRT > \angle PTR$. In $\triangle PRT$, $PT > RP$

9. **(C)**
$$x^5 = 167$$
$$x = (167)^{\frac{1}{5}}$$
$$= 2.783$$

10. **(A)** From $\triangle FGH$, $GH = 5 \sin r$.
Since $\angle FGK$ is a right angle, $\angle HGK = r$.
In $\triangle HGK$, $HK = GH \sin r$ or $HK = 5 \sin^2 r$

11. **(D)** The roots of $2x^2 - 6x + C = 0$ are equal and the discriminant is equal to 0,
$$b^2 - 4ac = 0$$
$$36 - 4 \cdot 2 \cdot C = 0$$
$$36 = 8C$$
$$C = 4\frac{1}{2}$$

12. **(A)**

$$\sqrt[3]{2x+4} = -.375$$
$$2x+4 = (-.375)^3$$
$$2x = -4.0527$$
$$x = -2.026$$

13. **(E)**

$$T = 2\pi\sqrt{\frac{L}{g}}$$

Squaring, we obtain $T^2 = 4\pi^2 \cdot \frac{L}{g}$

$$\frac{gT^2}{4\pi^2} = L$$

14. **(B)** The locus of points 7″ from P is a sphere of radius 7″. The locus of points 5″ from m consists of two planes above and below m. The sphere intersects only the upper plane in a *circle*.

15. **(D)** The right branch of the graph has slope 1 and y-intercept of 1.
Hence, its equation is $y = x + 1$.
To the left of $x = -1$, this line, $y = x + 1$, continues below the y-axis. We reflect it above the x-axis by making the equation $y = |x+1|$.

16. **(A)** We first define natural numbers, then integers, to include negative numbers, then rational numbers, and then imaginary numbers.

17. **(E)**

$$\frac{1}{y-1} = y+1$$
$$1 = (y+1)(y-1) = y^2 - 1$$
$$2 = y^2$$
$$y = \pm\sqrt{2}$$

18. **(C)**

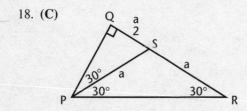

Let $RS = a$. Since PSR is an isosceles $\triangle$, $PS = a$.
In right $\triangle PQS$, $QS = \frac{a}{2}$

$$\frac{QS}{SR} = \frac{1}{2}$$

19. **(A)** $4y^2 - 3y + C = 0$
Since the roots are real,

$$9 - 16C \geq 0$$
$$\text{or } 9 \geq 16C$$
$$C \leq \frac{9}{16}$$

The product of the roots is $\frac{C}{4}$, and this is a maximum when $C = \frac{9}{16}$.

Solutions

20. **(B)** $2 + 4 + 6 + \cdots + 50$

$$S = \frac{n}{2}(a+1)$$
$$= \frac{25}{2}(2+50)$$
$$= \frac{25}{2} \times 52 = 650$$

21. **(D)** If $\frac{a}{b} = \frac{c}{d}$, by cross-multiplying, $ad = bc$. Therefore it is not possible for $\frac{a}{d} = \frac{b}{c}$.

22. **(A)**
$$(x+2)^2 + (y+3)^2 = (x+2)^2 + (y-5)^2$$
$$y^2 + 6y + 9 = y^2 - 10y + 25$$
$$16y = 16$$
$$y = 1$$

23. **(C)**
$$f(x) = \frac{x+1}{x-1}$$
$$f\left(\frac{3}{5}\right) = -4$$
$$f(-4) = .6$$
$$f(.6) = -4$$
$$f(-4) = .6$$

24. **(B)**

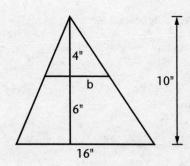

$$\frac{b}{4} = \frac{16}{10}$$
$$b = \frac{64}{10} = 6.4$$
$$A = \frac{1}{2}h(b+b')$$
$$A = \frac{1}{2} \cdot 6(16 + 64)$$
$$= 3 \times 22.4$$
$$= 67.2 T^2$$

25. **(C)** The locus of the centers is a circle with *P* as center and *r* as radius.

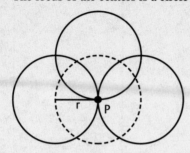

26. **(C)** The graph of $x^2 + y^2 = 4$ is a circle of radius 2 with center at the origin. The graph of $y^2 = 4$ consists of two horizontal lines, $y = +2$ and $y = -2$. These lines are tangent to the circle at $(0, 2)$ and $(0, -2)$. There are two points in common.

27. **(E)** If "at least one senior is an immature student," it is false that "all seniors are mature students."

28. **(B)** Tangent to the x-axis means the roots are real and equal.
Therefore, $b^2 - 4ac = 0$
$a = 2.8$, $b = -\sqrt{5}$, $c = c$

$$\left(-\sqrt{5}\right)^2 - 4(2.8)c = 0$$
$$11.2c = 5$$
$$c = \frac{5}{11.2} = .4464$$

29. **(D)** If $y \geq 5$, $y - 5 < 6$ and $y < 11$
If $y < 5$, $5 - y < 6$, $-y < 1$ and $y > -1$
The set of values is $-1 < y < 11$.

30. **(A)** If we subtract $\frac{5}{r-1}$ from both sides of the equation, it appears that $r = 1$.
But $\frac{5}{r-1}$ is not defined for $r = 1$. Hence, there is no root.

31. **(C)** Let *x* quarts = capacity.

$$\frac{1}{2}x + 6 = \frac{2}{3}x$$
$$3x + 36 = 4x$$
$$36 = x$$

32. **(C)** Let N = no. of revolutions made by the larger wheel.

$$\frac{10}{14} = \frac{N}{N+50} = \frac{5}{7}$$

distance = 125 × circumference
= 125 × 14π
= 1750π

$7N = 5N + 250$
$2N = 250$
$N = 125$

33. **(C)** These graphs have the same slope but different y-intercept. Hence, the graphs of the equations are parallel lines.

34. **(B)**

$$x^2 + (3.14)^2 = (6.2)^2$$

$$\Downarrow$$

$$x^2 = 38.44 - 9.86$$
$$x = 5.346$$

35. **(E)** The radius of O' is one-half that of O. Therefore, the area of circle O' is $\frac{1}{4}$ that of O. The area of O' is $\frac{1}{4}$ of 16, or 4.

36. **(B)**

$$x^2 + y^2 = 26^2, \quad x - y = 14 \text{ or } x = 14 + y$$
$$(14 + y)^2 + y^2 = 26^2$$
$$14^2 + 28y + y^2 + y^2 = 26^2$$
$$2y^2 + 28y + 196 = 676$$
$$y^2 + 14y + 98 = 338$$
$$y^2 + 14y - 240 = 0$$
$$(y + 24)(y - 10) = 0$$

$$y = -24 \text{ or } y = 10$$

Alternate solution:

$$x^2 + y^2 = 26^2$$
$$(x - y)^2 = 14^2$$
$$x^2 + y^2 - 2xy = 14^2$$

Substitute $x^2 + y^2 = 26^2$

$$26^2 - 2xy = 14^2$$
$$2xy = 26^2 - 14^2$$
$$2xy = 676 - 196 = 480$$
$$(x + y)^2 = x^2 + 2xy + y^2$$
$$= 26^2 + 2xy$$
$$= 676 + 2xy$$

Substitute $2xy = 480$ from above.

$$= 676 + 480$$
$$= 1156$$
$$x + y = \sqrt{1156} = 34$$

37. **(C)** 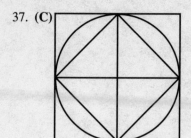 The smaller square is made up of 4 congruent triangles, and the larger square is made up of 8 congruent triangles. The ratio of their areas is $1:2$.

38. **(B)**
$$f(x) = x - \frac{1}{x}$$
$$f\left(\frac{1}{x}\right) = \frac{1}{x} - x$$
$$f(-x) = -x + \frac{1}{x} = \frac{1}{x} - x$$

Hence, $f\left(\frac{1}{x}\right) = f(-x)$

Also, $-f(x) = -x + \frac{1}{x} = \frac{1}{x} - x$

$f\left(\frac{1}{x}\right)$ = II and III

39. **(C)** $\dfrac{3\sqrt{18}}{2\sqrt{6}} = \dfrac{3}{2}\sqrt{\dfrac{18}{6}} = \dfrac{3}{2}\sqrt{3}$ which is an irrational number. The other choices are all rational.

40. **(B)**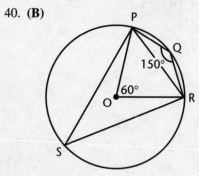

Minor $\widehat{PQR} = L$
If $\widehat{PSR} = 300°$, $\widehat{PQR} = 60°$
$$L = \frac{60}{360} \times 2\pi 10$$
$$= \frac{1}{6} \times 20\pi = \frac{10\pi}{3}$$
$$= \frac{10 \times 3.14}{3} \approx 10.4$$
$10 < L < 10.5$

41. **(A)**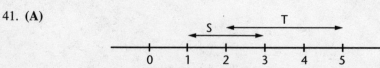

The set $S \cap T$ consists of all real numbers x such that $2 \leq x \leq 3$.

42. **(B)**
$$x^2 - 3x - 7 = 0$$
$$a = 1,\ b = -3,\ c = -7$$
$$x = \frac{-b \pm \sqrt{b^2 - 4ac}}{2a}$$
$$= \frac{-(-3) \pm \sqrt{(-3)^2 - 4(1)(-7)}}{2(1)}$$
$$= \frac{3 \pm \sqrt{37}}{2}$$
$$\frac{3 + \sqrt{37}}{2} = 4.54 \qquad \frac{3 - \sqrt{37}}{2} = -1.54$$

43. **(D)**
$$\% \text{ increase} = \frac{\text{increase}}{\text{original}} \times 100$$
$$= \frac{y - x}{x} \times 100 = \frac{100(y - x)}{x}$$

44. **(B)** For any point on line OQ, the abscissa equals the ordinate. Since P is to the right of the line $x = y$, it follows that $x > y$.

45. **(C)**

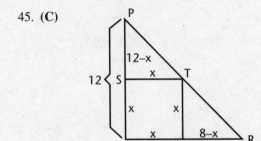

Since $\triangle PST \sim \triangle TVR$,
$$\frac{12 - x}{x} = \frac{x}{8 - x}$$
$$x^2 = (12 - x)(8 - x)$$
$$x^2 = 96 - 20x + x^2$$
$$20x = 96$$
$$x = 4.8$$

46. **(A)** $\quad \dfrac{S}{S'} = \dfrac{2^2}{3^2} = \dfrac{4}{9}$

47. **(B)** $\quad \dfrac{V}{V'} = \dfrac{2^3}{3^3} = \dfrac{8}{27}$

48. **(C)** $\quad \dfrac{12 \cdot 2}{12 \cdot 3} = \dfrac{2}{3}$

49. **(C)** $\quad \dfrac{d}{d'} = \dfrac{2\sqrt{2}}{3\sqrt{2}} = \dfrac{2}{3}$

50. **(E)** $\quad \dfrac{D}{d} = \dfrac{2\sqrt{3}}{2\sqrt{2}} = \dfrac{\sqrt{3}}{\sqrt{2}}$

Sample Test 4
Answer Sheet

Math Level IC

1. Ⓐ Ⓑ Ⓒ Ⓓ Ⓔ
2. Ⓐ Ⓑ Ⓒ Ⓓ Ⓔ
3. Ⓐ Ⓑ Ⓒ Ⓓ Ⓔ
4. Ⓐ Ⓑ Ⓒ Ⓓ Ⓔ
5. Ⓐ Ⓑ Ⓒ Ⓓ Ⓔ
6. Ⓐ Ⓑ Ⓒ Ⓓ Ⓔ
7. Ⓐ Ⓑ Ⓒ Ⓓ Ⓔ
8. Ⓐ Ⓑ Ⓒ Ⓓ Ⓔ
9. Ⓐ Ⓑ Ⓒ Ⓓ Ⓔ
10. Ⓐ Ⓑ Ⓒ Ⓓ Ⓔ
11. Ⓐ Ⓑ Ⓒ Ⓓ Ⓔ
12. Ⓐ Ⓑ Ⓒ Ⓓ Ⓔ
13. Ⓐ Ⓑ Ⓒ Ⓓ Ⓔ
14. Ⓐ Ⓑ Ⓒ Ⓓ Ⓔ
15. Ⓐ Ⓑ Ⓒ Ⓓ Ⓔ
16. Ⓐ Ⓑ Ⓒ Ⓓ Ⓔ
17. Ⓐ Ⓑ Ⓒ Ⓓ Ⓔ
18. Ⓐ Ⓑ Ⓒ Ⓓ Ⓔ
19. Ⓐ Ⓑ Ⓒ Ⓓ Ⓔ
20. Ⓐ Ⓑ Ⓒ Ⓓ Ⓔ
21. Ⓐ Ⓑ Ⓒ Ⓓ Ⓔ
22. Ⓐ Ⓑ Ⓒ Ⓓ Ⓔ
23. Ⓐ Ⓑ Ⓒ Ⓓ Ⓔ
24. Ⓐ Ⓑ Ⓒ Ⓓ Ⓔ
25. Ⓐ Ⓑ Ⓒ Ⓓ Ⓔ
26. Ⓐ Ⓑ Ⓒ Ⓓ Ⓔ
27. Ⓐ Ⓑ Ⓒ Ⓓ Ⓔ
28. Ⓐ Ⓑ Ⓒ Ⓓ Ⓔ
29. Ⓐ Ⓑ Ⓒ Ⓓ Ⓔ
30. Ⓐ Ⓑ Ⓒ Ⓓ Ⓔ
31. Ⓐ Ⓑ Ⓒ Ⓓ Ⓔ
32. Ⓐ Ⓑ Ⓒ Ⓓ Ⓔ
33. Ⓐ Ⓑ Ⓒ Ⓓ Ⓔ
34. Ⓐ Ⓑ Ⓒ Ⓓ Ⓔ
35. Ⓐ Ⓑ Ⓒ Ⓓ Ⓔ
36. Ⓐ Ⓑ Ⓒ Ⓓ Ⓔ
37. Ⓐ Ⓑ Ⓒ Ⓓ Ⓔ
38. Ⓐ Ⓑ Ⓒ Ⓓ Ⓔ
39. Ⓐ Ⓑ Ⓒ Ⓓ Ⓔ
40. Ⓐ Ⓑ Ⓒ Ⓓ Ⓔ
41. Ⓐ Ⓑ Ⓒ Ⓓ Ⓔ
42. Ⓐ Ⓑ Ⓒ Ⓓ Ⓔ
43. Ⓐ Ⓑ Ⓒ Ⓓ Ⓔ
44. Ⓐ Ⓑ Ⓒ Ⓓ Ⓔ
45. Ⓐ Ⓑ Ⓒ Ⓓ Ⓔ
46. Ⓐ Ⓑ Ⓒ Ⓓ Ⓔ
47. Ⓐ Ⓑ Ⓒ Ⓓ Ⓔ
48. Ⓐ Ⓑ Ⓒ Ⓓ Ⓔ
49. Ⓐ Ⓑ Ⓒ Ⓓ Ⓔ
50. Ⓐ Ⓑ Ⓒ Ⓓ Ⓔ

Directions: For each question in the sample test, select the best of the answer choices and blacken the corresponding space on this answer sheet.

Please note: (a) You will need to use a calculator in order to answer some, though not all, of the questions in this test. As you look at each question, you must decide whether or not you need a calculator for the specific question. A four-function calculator is not sufficient; your calculator must be at least a scientific calculator. Calculators that can display graphs and programmable calculators are also permitted.

(b) The only angle measure used on the Level IC test is degree measure. Your calculator should be set to degree mode.

(c) All figures are accurately drawn and are intended to supply useful information for solving the problems that they accompany. Figures are drawn to scale UNLESS it is specifically stated that a figure is not drawn to scale. Unless otherwise indicated, all figures lie in a plane.

(d) The domain of any function f is assumed to be the set of all real numbers x for which $f(x)$ is a real number except when this is specified not to be the case.

(e) Use the reference data below as needed.

REFERENCE DATA

Solid	Volume	Other
Right circular cone	$V = \frac{1}{3}\pi r^2 h \quad S = cl$	V = volume, S = lateral area, r = radius, h = height, c = circumference of base, l = slant height
Sphere	$V = \frac{4}{3}\pi r^3 \quad S = 4\pi r^2$	V = volume, r = radius, S = surface area
Pyramid	$V = \frac{1}{3}Bh$	V = volume, B = area of base, h = height

Sample Test 4

MATH LEVEL IC

50 Questions • Time—60 Minutes

DO YOUR FIGURING HERE.

1. The equation $y + \sqrt{y-4} = 6$ has

   (A) two real roots
   (B) only one real root
   (C) no real roots
   (D) one real root and one imaginary root
   (E) only one imaginary root

2. If $s = \dfrac{at}{a+t}$, $t =$

   (A) $\dfrac{as}{a-s}$ (B) $\dfrac{as}{a+s}$ (C) $\dfrac{a-s}{as}$ (D) $\dfrac{a+s}{as}$ (E) none of these

3. In figure 3, what is the area of triangle *NJL*?

   (A) 3.5
   (B) 53.72
   (C) 57.72
   (D) 80.53
   (E) 115.44

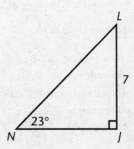

Fig. 3

4. The set of points in space 4 inches from a given line and 4 inches from a given point on this line is

   (A) a set consisting of two points
   (B) a set consisting of four points
   (C) a set consisting of two circles
   (D) the empty set
   (E) a circle

5. The multiplicative inverse of $-1+\sqrt{2}$ is

   (A) $-1-\sqrt{2}$ (B) $1-\sqrt{2}$ (C) $1+\sqrt{2}$ (D) $\sqrt{2}$
   (E) $2-\sqrt{2}$

6. Given equilateral triangle ABC of side 5. The midpoints of the sides are joined to form $\triangle DEF$; the midpoints of $\triangle DEF$ are then joined to form $\triangle GHI$, and this process is continued infinitely. The sum of the perimeters of all the triangles formed is

   (A) 30 (B) 25 (C) 35 (D) 38 (E) 40

7. If the graphs of the equations $2x + 5y = 7$ and $6x + cy = 11$ are parallel lines, c is equal to

   (A) 3 (B) 5 (C) 10 (D) 12 (E) 15

8. On the line $y = 2x - 1$, what is the distance between the points where $x = 1$ and $x = 2$?

   (A) 2.24
   (B) 2.27
   (C) 2.31
   (D) 2.42
   (E) 2.45

9. In the formula $s = \frac{1}{2}gt^2$, where g is a constant, the effect of tripling the value of t is to

   (A) triple the value of s
   (B) multiply the value of s by 9
   (C) multiply the value of s by 6
   (D) multiply the value of s by 36
   (E) multiply the value of s by 9g

10. In a triangle with sides of 7 and 9, the third side must be

    (A) more than 16
    (B) between 7 and 9
    (C) between 2 and 16
    (D) between 7 and 16
    (E) between 9 and 16

DO YOUR FIGURING HERE.

11. Lines p and q are two parallel lines in space. How many planes may be drawn that contain p and are parallel to q?

    (A) none
    (B) one
    (C) two
    (D) three
    (E) an infinite number

12. In $\triangle PQR$, $\angle P = 70°$, QS bisects $\angle Q$ and RS bisects $\angle R$. $\angle S =$

    (A) 110°
    (B) 115°
    (C) 120°
    (D) 125°
    (E) 130°

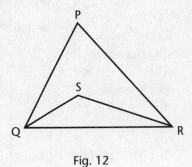

Fig. 12

13. A rectangular garden with dimensions u and v is bordered by a walk of uniform width w. The area of the walk is

    (A) $4w^2$
    (B) $2w(u + v)$
    (C) $(u + w)(v + w)$
    (D) $(u + w)(v + w) - uv$
    (E) $(u + 2w)(v + 2w) - uv$

14. In figure 14, $RS \perp ST$, $PT = 40$, and $PR = RS = 15$. $ST =$

    (A) 10
    (B) 15
    (C) 20
    (D) 25
    (E) 28

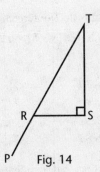

Fig. 14

15. A right circular cylinder of height h is inscribed in a cube of height h so that the bases of the cylinder are inscribed in the upper and lower faces of the cube. The ratio of the volume of the cylinder to that of the cube is

    (A) $\pi : 4$ (B) $4 : \pi$ (C) $3 : 4$ (D) $2 : 3$ (E) $\pi : 1$

GO ON TO THE NEXT PAGE

16. In figure 16, altitude *RH* = 15 and *ST* is drawn parallel to *QP*. What must be the length of *RJ* so that $\triangle RST = \frac{1}{3} \triangle RQP$?

    (A) 5
    (B) $5\sqrt{3}$
    (C) $5\sqrt{2}$
    (D) 7
    (E) cannot be determined from the information given

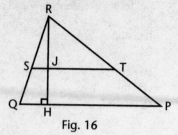

Fig. 16

17. If $p - q > 0$, which of the following is true?

    (A) If $q = 0, p < 0$
    (B) If $q < 0, p < 0$
    (C) If $p > 0, q > 0$
    (D) If $q = 0, p > 0$
    (E) If $q < 1, p > 1$

18. In figure 18, chords *PQ* and *RS* intersect at *T*. If $\angle R = 50°$ and $\angle P = 46°$, the number of degrees in minor arc *PR* is

    (A) 84
    (B) 168
    (C) 42
    (D) 130
    (E) cannot be determined from the information given

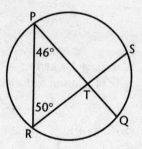

Fig. 18

19. If $f(x) = 3x - 2$ and $g(x) = 7$; $f[g(x)] =$

    (A) $21x - 2$ (B) 7   (C) 19   (D) $7x - 2$ (E) 13

20. What is the *y*-intercept of the line whose equation is
    $x + 3 - \sqrt{3}(y - 5) = 0$?

    (A) 6.43
    (B) 6.53
    (C) 6.63
    (D) 6.73
    (E) 6.83

21. What is the circumference of a circle whose area is 7.2?

    (A) 9.51
    (B) 9.15
    (C) 7.20
    (D) 4.13
    (E) 3.14

22. In △STU, ∠T = 70°, ST = 4 and TU = 5. Which of the following conclusions may be drawn?

    (A) ∠U = 55°
    (B) ∠S > 55°
    (C) SU = 6
    (D) ∠U > 55°
    (E) ∠S : ∠U = 5 : 4

DO YOUR FIGURING HERE.

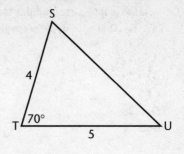

Fig. 22

23. If d varies directly as the square of t and if $d = 18$ when $t = 3$, the value of d when $t = 5$ is

    (A) 40   (B) 45   (C) 50   (D) 55   (E) 60

24. If the graphs of the equations $x^2 + y^2 = 4$ and $x^2 - y^2 = 9$ are drawn on the same set of axes, the number of intersections of the curves is

    (A) 4   (B) 3   (C) 2   (D) 1   (E) 0

25. If $3^{n-2} = \frac{1}{81}$, $n =$

    (A) 0   (B) 1   (C) 2   (D) –2   (E) –4

26. What is the radius of the circle whose equation is $(x - \sqrt{3})^2 + (y + 2)^2 = 11$?

    (A) 1.71
    (B) 2.33
    (C) 3.32
    (D) 3.85
    (E) 4.27

GO ON TO THE NEXT PAGE

27. Water is poured from a full right circular cylinder of height 3 ft and radius 2 ft into a rectangular tank whose base dimensions are 7 ft by 1 ft. To what height, in feet, will the water rise (assume no overflow)?

   (A) 5.49
   (B) 5.39
   (C) 5.34
   (D) 5.29
   (E) 5.24

28. If $f(x) = 5x - 3$ and $f(t) = 7$, $t =$

   (A) 0   (B) 1   (C) 2   (D) 32   (E) −2

29. If the fraction $\frac{a}{b} > 2$ where a and b are positive, then

   (A) $2a = 2b$ (B) $a > b$ (C) $a > 2$ (D) $a > 2b$ (E) $2b > a$

30. The ratio of the sides of 2 cubes is 3:4. The difference of their volumes is 296. The side of the smaller cube is

   (A) 8   (B) 9   (C) 3   (D) 4   (E) 6

31. How many numbers in the set $\{-6, -3, 0, 3, 6\}$ satisfy the inequality $|2x - 4| < 11$?

   (A) 0   (B) 1   (C) 3   (D) 4   (E) 5

32. How many multiples of 3 are between 23 and 82?

   (A) 19   (B) 20   (C) 21   (D) 22   (E) 23

33. The graph of the equation $x^2 + \frac{y^2}{4} = 1$ is

   (A) an ellipse
   (B) a circle
   (C) a hyperbola
   (D) a parabola
   (E) two straight lines

34. If $\cos x = m$ and $0 < x < 90°$, $\tan x =$

   (A) $\dfrac{\sqrt{1-m^2}}{m}$ (B) $\dfrac{m}{\sqrt{1-m^2}}$ (C) $\dfrac{\sqrt{1+m^2}}{m}$ (D) $\dfrac{m}{\sqrt{1+m^2}}$
   (E) $\dfrac{1-m^2}{m}$

35. Which pair of points in figure 35 must be joined to produce a line with slope of -1?

   (A) P and Q
   (B) R and S
   (C) P and R
   (D) Q and S
   (E) Q and R

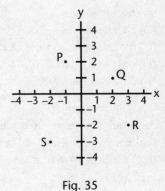

Fig. 35

36. If $f(t) = \dfrac{1+t}{1-t}$, $f(-t) =$

   (A) $\dfrac{1}{f(t)}$ (B) $\dfrac{1}{f(-t)}$ (C) $f\left(\dfrac{1}{t}\right)$ (D) $f\left(-\dfrac{1}{t}\right)$ (E) $1 + f(t)$

37. Let x be the angle formed by the diagonal of the face of a cube and the diagonal of the cube, drawn from the same vertex. $\cos^2 x =$

   (A) $\dfrac{1}{3}$ (B) $\dfrac{1}{2}$ (C) $\dfrac{3}{4}$ (D) $\dfrac{3}{2}$ (E) $\dfrac{2}{3}$

38. The vertices of a triangle are $(2, 0)$, $(-2, -1)$, and $(3, -4)$. The triangle is

   (A) right scalene
   (B) isosceles right
   (C) equilateral
   (D) scalene but not right
   (E) none of these

39. An illustration of the associative law for multiplication is given by

   (A) $\left(\dfrac{1}{3} \times 5\right) \times 8 = \dfrac{1}{3} \times (5 \times 8)$
   (B) $\dfrac{1}{3} \times 5 \times 8 = \dfrac{1}{3} \times 8 \times 5$
   (C) $\dfrac{1}{3} \times 5 + \dfrac{1}{3} \times 8 = \dfrac{1}{3} \times 13$
   (D) $\dfrac{1}{3} \times 5 \times 8 = \left(\dfrac{1}{3} \times 5\right) \times \left(\dfrac{1}{3} \times 8\right)$
   (E) none of these

40. The graph of the equation $x^2 + y^2 = 169$ is a circle. Which of the following points is outside the circle?

   (A) (5, 12)   (B) (9, 9)   (C) (11, 5)   (D) (10, 10)   (E) none of these

41. The statement $\sin x - \cos x = 0$ is true for

   (A) no values of x
   (B) all values of x
   (C) more than two values of x, but not all values
   (D) only one value of x
   (E) only two values of x

42. A rectangular piece of cardboard is 40 in. wide and 50 in. long. Squares 5 in. on a side are cut out of each corner, and the remaining flaps are bent up to form an open box. The number of cubic inches in the box is

   (A) 1,200   (B) 7,875   (C) 6,000   (D) 8,000   (E) 10,000

43. If $p - q = q - p$, then

   (A) $p - q = 1$ (B) $p + q = 1$ (C) $p = 2q$ (D) $pq = 1$
   (E) $\dfrac{p}{q} = 1$

44. Let $\boxed{n}$ be defined as $\dfrac{(n+2)!}{(n-1)!}$. What is the value of $\dfrac{\boxed{7}}{\boxed{3}}$?

   (A) 4.4
   (B) 8.4
   (C) 12.4
   (D) 16.4
   (E) 20.4

45. What is the value of x for the following system of equations?

$$\sqrt{2}x - \sqrt{3}y = \sqrt{5}$$
$$2x + 3y = 5$$

   (A) 1.59
   (B) 1.69
   (C) 1.79
   (D) 1.89
   (E) 1.99

46. To prove $x > y$, we may first show that the suppositions $x < y$ and $x = y$ lead to conclusions that contradict given conditions. Such a proof is called

    (A) circular reasoning
    (B) proof by induction
    (C) indirect proof
    (D) reasoning from a converse
    (E) proof by analogy

47. R is the set of all positive odd integers less than 20; S is the set of all multiples of 3 that are less than 20. How many elements in the set
    $R \cap S$?

    (A) 0   (B) 1   (C) 2   (D) 3   (E) 4

48. If $3^{2x} = 200$, x is between

    (A) 0 and 1
    (B) 1 and 2
    (C) 2 and 3
    (D) 3 and 4
    (E) 4 and 5

49. If the graph of $y = x^2 - 3x + K$ is tangent to the x-axis, the roots of $x^2 - 3x + K$ could be

    (A) imaginary
    (B) real, equal, and rational
    (C) real, unequal, and irrational
    (D) real, unequal, and rational
    (E) not determinable from the information given

50. If $\log_x 81 = 4$, $x =$

    (A) 1   (B) 2   (C) 3   (D) 4   (E) 5

STOP

END OF SAMPLE TEST 4. IF YOU HAVE ANY TIME LEFT, GO OVER YOUR WORK IN THIS SECTION ONLY. DO NOT WORK IN ANY OTHER SECTION OF THE TEST.

Sample Test 4
Answer Key

Math Level IC

1. B	11. E	21. A	31. D	41. C
2. A	12. D	22. B	32. B	42. C
3. C	13. E	23. C	33. A	43. E
4. E	14. C	24. E	34. A	44. B
5. C	15. A	25. D	35. C	45. E
6. A	16. B	26. C	36. A	46. C
7. E	17. D	27. B	37. E	47. D
8. A	18. E	28. C	38. B	48. C
9. B	19. C	29. D	39. A	49. B
10. C	20. D	30. E	40. D	50. C

Solutions

1. **(B)**

$$y + \sqrt{y-4} = 6$$
$$\sqrt{y-4} = 6 - y$$
$$y - 4 = 36 - 12y + y^2$$
$$y^2 - 13y + 40 = 0$$
$$(y-8)(y-5) = 0$$
$$y = 8, \quad y = 5$$

Check $y = 8$.

$8 + \sqrt{4} = 6$
$8 + 2 = 6$ (no check)

Check $y = 5$.

$5 + \sqrt{1} = 6$ (check)

Hence, there is only one real root; $y = 5$

2. **(A)**

$$s = \frac{at}{a+t}$$
$$sa + st = at$$
$$sa = at - st$$
$$sa = t(a-s)$$
$$t = \frac{as}{s-s}$$

208

3. **(C)**

$\tan 23° = \dfrac{7}{x}$

$x = \dfrac{7}{\tan 23°} = 16.491$

$A = \dfrac{1}{2}bh$

$= \dfrac{1}{2}(16.491)(7)$

$= 57.72$

4. **(E)** The locus of points 4" from a given line is a cylindrical surface of radius 4" with the given line as axis. The locus of points 4" from a given point is a sphere of 4" radius with the given point as center. The intersection of these two loci is a circle.

5. **(C)** $\dfrac{1}{-1+\sqrt{2}} \cdot \dfrac{-1-\sqrt{2}}{-1-\sqrt{2}} = \dfrac{-1-\sqrt{2}}{1-2} = 1+\sqrt{2}$

6. **(A)** The sum of the perimeters is

$$S = 15 + \dfrac{15}{2} + \dfrac{15}{4} + \dfrac{15}{8} + \ldots$$

This is an infinite geometric progression with $a = 15$ and $r = \dfrac{1}{2}$.

$$S = \dfrac{a}{1-r}$$

$$S = \dfrac{15}{1-\dfrac{1}{2}}$$

$$S = \dfrac{15}{\dfrac{1}{2}}$$

$$S = 30$$

7. **(E)** If the lines are parallel, they must have the same slopes. Since $6 = 3 \cdot 2$, $C = 3 \cdot 5 = 15$.

8. **(A)**

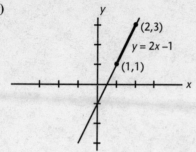

$$d = \sqrt{(x_2 - x_1)^2 + (y_2 - y_1)^2}$$
$$= \sqrt{(2-1)^2 + (3-1)^2}$$
$$= \sqrt{1+4} = \sqrt{5}$$
$$= 2.236$$

9. **(B)** Since t is squared in the formula $s = \frac{1}{2}gt^2$, the effect of tripling the value of t would be to multiply s by 9.

10. **(C)** If x is the third side, $x < 7 + 9$ or $x < 16$. Also $x > 9 - 7$ or $x > 2$. x must be between 2 and 16.

11. **(E)** If line p is parallel to line q, then any plane containing p will either contain q or be parallel to it. There is an *infinite* number of such planes.

12. **(D)** Since $\angle P = 70°$, $\angle PQR + \angle PRQ = 110°$. Since these latter two angles are bisected, $\angle SQR + \angle SRQ = 55°$. Therefore, in $\triangle QSR$, $\angle S = 180° - 55° = 125°$.

13. **(E)**

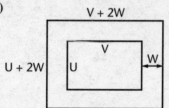

The area of the large rectangle is $(u + 2w)(v + 2w)$
The area of the small rectangle is uv.
The area of the walk is $(u + 2w)(v + 2w) - uv$.

14. **(C)** Since $PT = 40$ and $PR = 15$, $RT = 25$. Because $RS = 15$, $ST = 20$, since RST is a 3-4-5 right $\triangle$.

15. **(A)**

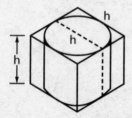

The diameter of each base of the cylinder is h.

Hence, the area of each base is $\frac{\pi h^2}{4}$.

The volume of the cylinder is $\frac{\pi h^2}{4} + h$.

The volume of the cube is h^3. Hence, the ratio of cylinder to cube is $\pi:4$.

16. **(B)** Let $RJ = x$.
    Since $\triangle RST \sim \triangle RQP$,
    $$\frac{\text{Area of } \triangle RST}{\text{Area of } \triangle RQP} = \frac{\overline{RJ}^2}{\overline{RH}^2}$$
    $$\frac{1}{3} = \frac{x^2}{15^2} \text{ or } \frac{x}{15} = \frac{1}{\sqrt{3}}$$
    $$x = \frac{15}{\sqrt{3}} \cdot \frac{\sqrt{3}}{\sqrt{3}} = 5\sqrt{3}$$

17. **(D)** If in $p - q > 0$ we let $q = 0$, it follows that $p - 0 > 0$ or $p > 0$.

18. **(E)** From the information given, we can determine any of the angles at T. However, since we do not know $\widehat{SQ}$ and cannot compute it, it follows that there is not enough information to compute $\widehat{PR}$.

19. **(C)** $f(x) = 3x - 2$, $g(x) = 7$
    $$f[g(x)] = 3[g(x)] - 2$$
    $$= 3 \cdot 7 - 2 = 19$$

20. **(D)** Method 1:
    $$x + 3 - \sqrt{3}(y - 5) = 0$$
    $$\Downarrow$$
    $$\sqrt{3}(y - 5) = x + 3$$
    $$y - 5 = \frac{x + 3}{\sqrt{3}}$$
    $$y = \frac{x + 3}{\sqrt{3}} + 5$$
    $$y = \frac{1}{\sqrt{3}}x + \left(\frac{3}{\sqrt{3}} + 5\right)$$
    the y-intercept is $\left(\frac{3}{\sqrt{3}} + 5\right) = 6.73$

    Method 2: To find the y-intercept, let $x = 0$
    $$x + 3 - \sqrt{3}(y - 5) = 0$$
    $$0 + 3 - \sqrt{3}(y - 5) = 0$$
    $$-\sqrt{3}y + 5\sqrt{3} = -3$$
    $$y = \frac{5\sqrt{3} + 3}{\sqrt{3}}$$
    $$= 6.73$$

21. **(A)**
$$A = \pi r^2 = 7.2$$
$$r = \sqrt{\frac{7.2}{\pi}} = 1.5139$$
$$C = 2\pi r = 2\pi(1.5139)$$
$$= 9.512$$

22. **(B)** If $\angle T = 70°$, $\angle S + \angle U = 110°$
Since $TU > ST$, $\angle S > \angle U$
$$\angle S > 55°.$$

23. **(C)**
$$d = Kt^2$$
$$18 = 9K$$
$$K = 2$$
$$d = 2t^2$$
When $t = 5, d = 2 \cdot 25$
$$d = 50$$

24. **(E)** The graph of $x^2 + y^2 = 4$ is a circle of radius 2 with center at the origin. The graph of $x^2 - y^2 = 9$ is a hyperbola with the x-axis as transverse axis and center at the origin; its x-intercepts are $(3, 0)$ and $(-3, 0)$. Hence, the curves have *no* intersections.

25. **(D)**
$$3^{n-2} = \frac{1}{81} = \frac{1}{3^4}$$
$$3^{n-2} = 3^{-4}$$
$$n - 2 = -4$$
$$n = -2$$

26. **(C)**
$$(x-h)^2 + (y-k)^2 = r^2$$
$$(x-\sqrt{3})^2 + (y+2)^2 = 11$$
Center: $(\sqrt{3}, -2)$
$$r = \sqrt{11} = 3.32$$

27. **(B)**
$$V_{\text{cylinder}} = \pi r^2 h = \pi(2)^2(3) = 12\pi$$

$$V_{\text{rectangular tank}} = l \cdot w \cdot h = 7\,(1)(x)$$

$$7(1)(x) = 12\pi$$

$$x = \frac{12\pi}{7} = 5.386 \text{ ft}$$

28. **(C)**
$$f(x) = 5x - 3 \text{ and } f(t) = 7$$
$$f(t) = 5t - 3 = 7$$
$$5t = 10$$
$$t = 2$$

29. **(D)** $\frac{a}{b} > 2$ and a and b are positive. Multiplying both sides by b, $a > 2b$

30. **(E)** Let the sides of the two cubes be represented by $3x$ and $4x$.

$$(4x)^3 - (3x)^3 = 296$$
$$64x^3 - 27x^3 = 296$$
$$37x^3 = 296$$
$$x^3 = 8$$
$$x = 2$$
$$3x = 6$$

31. **(D)** If $|2x - 3| < 11$,

$$-11 < 2x - 3 < 11$$
$$-8 < 2x < 14$$
$$-4 < x < 7$$

All numbers in the given set are in the interval except -6. Thus, *four* values in the set satisfy the inequality.

32. **(B)** 23, 24 ... 81, 82
The first multiple of 3 is 24 and the last 81. When divided by 3 these numbers give 8 and 27. Between 8 and 27 *inclusive* are 20 numbers. Hence, there are 20 multiples of 3.

33. **(A)** $\frac{x^2}{1} + \frac{y^2}{4} = 1$ graphs as an ellipse with major axis along the *y*-axis, a major axis 2 and minor axis of 1.

34. **(A)** If $\cos x = m$

$$\sin x = \sqrt{1 - \cos^2 x} = \sqrt{1 - m^2}$$

$$\tan x = \frac{\sin x}{\cos x} = \frac{\sqrt{1 - m^2}}{m}$$

35. **(C)** PQ, PR, and QR have negative slopes.

Slope of $PR = \dfrac{2-(-2)}{-1-3} = \dfrac{4}{-4} = -1$

PR has a slope of -1.

36. **(A)** $\quad f(t) = \dfrac{1+t}{1-t},\, f(-t) = \dfrac{1-t}{1+t}$

Thus $f(-t) = \dfrac{1}{f(t)}$

37. **(E)**

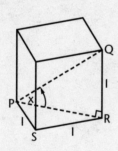

$$PQ = \sqrt{1^2 + 1^2 + 1^2} = \sqrt{3}$$
$$PR = \sqrt{1^2 + 1^2} = \sqrt{2}$$
$$\cos x = \frac{PR}{PQ} = \frac{\sqrt{2}}{\sqrt{3}}$$
$$\cos^2 x = \frac{2}{3}$$

38. **(B)** Vertices are $P(2, 0)$, $Q(-2, -1)$, and $R(3, -4)$.

$$PQ^2 = 4^2 + 1^2 = 17$$
$$PR^2 = 1^2 + 4^2 = 17$$
$$QR^2 = 5^2 + 3^2 = 34$$

Since $PQ = PR$, the triangle is isosceles.

Also, slope of $PQ = \dfrac{1}{4}$, slope of $QR = -4$. Thus $PQ \perp PR$.

39. **(A)** $\left(\dfrac{1}{3} \times 5\right) \times 8 = \dfrac{1}{3} \times (5 \times 8)$ illustrates the associative law $(a \times b) \times c = a \times (b \times c)$.

40. **(D)** The graph of $x^2 + y^2 = 169$ is a circle of radius 13 with center at the origin. Calculate the distance of each given point from the origin. For the point (10, 10), distance from origin is

$$\sqrt{10^2 + 10^2} = 10\sqrt{2} = 14.14$$

Since this distance is greater than 13, the point lies outside the circle.

41. **(C)** $\sin x - \cos x = 0$

$$\sin x = \cos x$$

$$\frac{\sin x}{\cos x} = 1 \quad \text{or} \quad \tan x = 1$$

$$x = 45°, 225°, 405°, \text{etc.}$$

There are many values.

42. **(C)**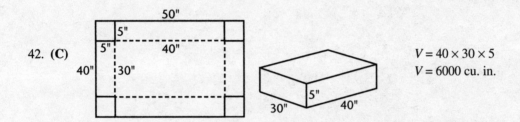

$V = 40 \times 30 \times 5$
$V = 6000$ cu. in.

43. **(E)**

$$p - q = q - p$$
$$2p = 2q$$
$$p = q$$
$$\frac{p}{q} = 1$$

44. **(B)**

$$\boxed{7} = \frac{9!}{6!} = 504$$

$$\boxed{3} = \frac{5!}{2!} = 60$$

$$\frac{504}{60} = 8.4$$

note: $\dfrac{(n+2)!}{(n-1)!} = \dfrac{(n+2)(n+1)(n)(n-1)(n-2)\dots(1)}{(n-1)(n-2)\dots(1)} = (n+2)(n+1)(n)$

216 SAT II: Math

45. **(E)** Method 1:
$$\sqrt{2}x - \sqrt{3}y = \sqrt{5}$$
$$2x + 3y = 5$$

To Eliminate y, multiply the first equation by 3 and the second equation by $\sqrt{3}$. Then add the equations.

$$3\sqrt{2}x - 3\sqrt{3}y = 3\sqrt{5}$$
$$2\sqrt{3}x + 3\sqrt{3}y = 5\sqrt{3}$$
$$(3\sqrt{2} + 2\sqrt{3})x = 3\sqrt{5} + 5\sqrt{3}$$

$$x = \frac{3\sqrt{5} + 5\sqrt{3}}{3\sqrt{2} + 2\sqrt{3}} = 1.99$$

Method 2:

$$x = \frac{\begin{vmatrix} \sqrt{5} & \sqrt{3} \\ 5 & 3 \end{vmatrix}}{\begin{vmatrix} \sqrt{2} & -\sqrt{3} \\ 2 & 3 \end{vmatrix}} = 1.99$$

46. **(C)** A proof of this type which disproves all possible alternatives to a desired conclusion is an *indirect proof*.

47. **(D)**
R{1, 3, 5, . . . 17, 19}
S{3, 6, 9, 12, . . . 15, 18}
R ∩ S = {3, 9, 15}   3 elements

48. **(C)** $3^{2x} = 200$

$$3^4 = 81 \quad \text{and} \quad 3^5 = 243$$

Therefore, $2x$ is between 4 and 5, or x is between 2 and 3.

49. **(B)** If the graph of a quadratic function, $y = f(x)$, is tangent to the x-axis, the roots of $f(x) = 0$ are real, equal, and rational.

50. **(C)**
$$\log_x 81 = 4$$
$$x^4 = 81$$
$$3^4 = 81$$
$$x = 3$$

PART SIX

Three Sample Mathematics Tests Level IIC

CONTENTS

Sample Test 1: Math Level IIC   221
Answer Key   228
Solutions   228
Sample Test 2: Math Level IIC   243
Answer Key   251
Solutions   251
Sample Test 3: Math Level IIC   263
Answer Key   271
Solutions   271

Sample Test 1
Answer Sheet

Math Level IIC

1 Ⓐ Ⓑ Ⓒ Ⓓ Ⓔ	11 Ⓐ Ⓑ Ⓒ Ⓓ Ⓔ	21 Ⓐ Ⓑ Ⓒ Ⓓ Ⓔ	31 Ⓐ Ⓑ Ⓒ Ⓓ Ⓔ	41 Ⓐ Ⓑ Ⓒ Ⓓ Ⓔ
2 Ⓐ Ⓑ Ⓒ Ⓓ Ⓔ	12 Ⓐ Ⓑ Ⓒ Ⓓ Ⓔ	22 Ⓐ Ⓑ Ⓒ Ⓓ Ⓔ	32 Ⓐ Ⓑ Ⓒ Ⓓ Ⓔ	42 Ⓐ Ⓑ Ⓒ Ⓓ Ⓔ
3 Ⓐ Ⓑ Ⓒ Ⓓ Ⓔ	13 Ⓐ Ⓑ Ⓒ Ⓓ Ⓔ	23 Ⓐ Ⓑ Ⓒ Ⓓ Ⓔ	33 Ⓐ Ⓑ Ⓒ Ⓓ Ⓔ	43 Ⓐ Ⓑ Ⓒ Ⓓ Ⓔ
4 Ⓐ Ⓑ Ⓒ Ⓓ Ⓔ	14 Ⓐ Ⓑ Ⓒ Ⓓ Ⓔ	24 Ⓐ Ⓑ Ⓒ Ⓓ Ⓔ	34 Ⓐ Ⓑ Ⓒ Ⓓ Ⓔ	44 Ⓐ Ⓑ Ⓒ Ⓓ Ⓔ
5 Ⓐ Ⓑ Ⓒ Ⓓ Ⓔ	15 Ⓐ Ⓑ Ⓒ Ⓓ Ⓔ	25 Ⓐ Ⓑ Ⓒ Ⓓ Ⓔ	35 Ⓐ Ⓑ Ⓒ Ⓓ Ⓔ	45 Ⓐ Ⓑ Ⓒ Ⓓ Ⓔ
6 Ⓐ Ⓑ Ⓒ Ⓓ Ⓔ	16 Ⓐ Ⓑ Ⓒ Ⓓ Ⓔ	26 Ⓐ Ⓑ Ⓒ Ⓓ Ⓔ	36 Ⓐ Ⓑ Ⓒ Ⓓ Ⓔ	46 Ⓐ Ⓑ Ⓒ Ⓓ Ⓔ
7 Ⓐ Ⓑ Ⓒ Ⓓ Ⓔ	17 Ⓐ Ⓑ Ⓒ Ⓓ Ⓔ	27 Ⓐ Ⓑ Ⓒ Ⓓ Ⓔ	37 Ⓐ Ⓑ Ⓒ Ⓓ Ⓔ	47 Ⓐ Ⓑ Ⓒ Ⓓ Ⓔ
8 Ⓐ Ⓑ Ⓒ Ⓓ Ⓔ	18 Ⓐ Ⓑ Ⓒ Ⓓ Ⓔ	28 Ⓐ Ⓑ Ⓒ Ⓓ Ⓔ	38 Ⓐ Ⓑ Ⓒ Ⓓ Ⓔ	48 Ⓐ Ⓑ Ⓒ Ⓓ Ⓔ
9 Ⓐ Ⓑ Ⓒ Ⓓ Ⓔ	19 Ⓐ Ⓑ Ⓒ Ⓓ Ⓔ	29 Ⓐ Ⓑ Ⓒ Ⓓ Ⓔ	39 Ⓐ Ⓑ Ⓒ Ⓓ Ⓔ	49 Ⓐ Ⓑ Ⓒ Ⓓ Ⓔ
10 Ⓐ Ⓑ Ⓒ Ⓓ Ⓔ	20 Ⓐ Ⓑ Ⓒ Ⓓ Ⓔ	30 Ⓐ Ⓑ Ⓒ Ⓓ Ⓔ	40 Ⓐ Ⓑ Ⓒ Ⓓ Ⓔ	50 Ⓐ Ⓑ Ⓒ Ⓓ Ⓔ

Directions: For each question in the sample test, select the best of the answer choices and blacken the corresponding space on this answer sheet.

Please note: (a) You will need to use a calculator in order to answer some, though not all, of the questions in this test. As you look at each question, you must decide whether or not you need a calculator for the specific question. A four-function calculator is not sufficient; your calculator must be at least a scientific calculator. Calculators that can display graphs and programmable calculators are also permitted.

(b) Set your calculator to radian mode or degree mode depending on the requirements of the question.

(c) All figures are accurately drawn and are intended to supply useful information for solving the problems that they accompany. Figures are drawn to scale UNLESS it is specifically stated that a figure is not drawn to scale. Unless otherwise indicated, all figures lie in a plane.

(d) The domain of any function f is assumed to be the set of all real numbers x for which $f(x)$ is a real number except when this is specified not to be the case.

(e) Use the reference data below as needed.

REFERENCE DATA

Solid	Volume	Other		
Right circular cone	$V = \frac{1}{3}\pi r^2 h$	$S = cl$	V = volume r = radius h = height	S = lateral area c = circumference of base l = slant height
Sphere	$V = \frac{4}{3}\pi r^3$	$S = 4\pi r^2$	V = volume r = radius S = surface area	
Pyramid	$V = \frac{1}{3}Bh$		V = volume B = area of base h = height	

TEAR HERE

Sample Test 1

MATH LEVEL IIC

50 Questions • Time—60 Minutes

DO YOUR FIGURING HERE.

1. The number of roots of the equation $9+\sqrt{x-3}=x$, is
   (A) 0   (B) 1   (C) 2   (D) 3   (E) ∞

2. The operation $\square$ is defined as $a \square b = a^b - b^a$. What is the value of $\left(\frac{1}{2}\right)^3 \square (3)^{\frac{1}{2}}$?

   A) 2.36   B) 1.93   C) .47   D) −.75
   E) −1.04

3. If $f(x) = 3x^2 - 5x - 4$ then $f(-2x)$ is equal to
   (A) $2f(-x)$ (B) $-f(x)$ (C) $4f(x)$ (D) $-4f(x)$
   (E) none of these

4. If $P = Ke^{-xt}$, then x equals

   (A) $\dfrac{\log K}{t \log e \log P}$ (B) $\dfrac{P}{Ke^t}$ (C) $\dfrac{Pe^t}{K}$

   (D) $\dfrac{\log K - \log P}{t \log e}$ (E) none of these

5. The vertices of a triangle are the intersections of the lines whose equations are $y = 0$, $x = 3y$, and $3x + y = 7$. This triangle is
   (A) isosceles   (B) equilateral   (C) right   (D) acute
   (E) obtuse

6. The area bounded by the closed curve whose equation is $x^2 - 6x + y^2 + 8y = 0$ is
   (A) 12π (B) 25π (C) 36π (D) 48π
   (E) cannot be determined

7. The ratio of the diagonal of a cube to the diagonal of a face of the cube is
   (A) $2:\sqrt{3}$ (B) $3:\sqrt{6}$ (C) $3:\sqrt{2}$ (D) $\sqrt{3}:1$
   (E) $\sqrt{6}:3$

8. A regular octagon is inscribed in a circle of radius 1. Find a side of the octagon.
   (A) $\sqrt{2}$ (B) $\dfrac{\sqrt{3}}{2}$ (C) $\sqrt{2+\sqrt{2}}$
   (D) $\sqrt{2-\sqrt{2}}$ (E) none of these

GO ON TO THE NEXT PAGE

9. Two circles of radii 3 inches and 6 inches have their centers 15 inches apart. Find the length in inches of the common internal tangent.

   (A) 8"   (B) 10"   (C) 12"   (D) 14"   (E) 15"

10. The graph of the equation $y = 5 \cos 3x$ has a period, in radians, of

    (A) $\frac{2\pi}{3}$ (B) $\frac{2\pi}{5}$ (C) 3π (D) 5   (E) 4

11. If $2^x = 8^{y+1}$ and $9^y = 3^{x-9}$ then y equals

    (A) 3   (B) 6   (C) 9   (D) 12   (E) 21

12. Express in terms of an inverse function the angle formed by the diagonals of a cube.

    (A) $\sin^{-1} 2/3$ (B) $\cos^{-1} 2/3$ (C) $\tan^{-1} 1/3$
    (D) $\sin^{-1} 1/3$ (E) $\cos^{-1} 1/3$

13. If $y = \dfrac{10^{\log x}}{x^2}$, for $x > 0$, then

    (A) y varies directly with x
    (B) y is independent of x
    (C) y varies as the square of x
    (D) $(xy)^2 = 3$
    (E) y varies inversely with x

14. If $\log_r 6 = m$ and $\log_r 3 = n$, then $\log_r\left(\dfrac{r}{2}\right)$ is equal to

    (A) $\dfrac{1}{2}\log_2 r$ (B) $1 - m - n$ (C) $1 - \log_r 2$
    (D) $\dfrac{r}{2}$ (E) $1 - m + n$

15. The inequality $-x^2 + x - 10 < -2x^2 - 4$ is satisfied if

    (A) $x < -3$ (B) $|x| < 3$ (C) $-3 < x < 2$
    (D) $-2 < x < 3$ (E) $x < -3$ or $x > 2$

16. The contrapositive of the sentence $\sim p \rightarrow q$ is equivalent to

    (A) $p \rightarrow \sim q$ (B) $q \rightarrow \sim p$ (C) $q \rightarrow p$
    (D) $\sim p \rightarrow \sim q$ (E) $\sim q \rightarrow p$

17. A point moves so that its distance from the origin is always twice its distance from the point $(3, 0)$. Its locus is

    (A) a circle   (B) an ellipse   (C) a hyperbola
    (D) a straight line   (E) a parabola

18. The function f is defined as $f = \{(x, y) | y = \dfrac{2x+1}{x-3}$ where $x \neq 3\}$. Find the value of K so that the inverse of f will be

    $f^{-1} = \{(x, y) | y = \dfrac{3x+1}{x-K}$ where $x \neq K\}$.

    (A) 1   (B) 2   (C) 3   (D) 4   (E) 5

19. Find the sum of the reciprocals of the roots of the equation $x^2 + px + q = 0$.

    (A) $-\dfrac{p}{q}$ (B) $\dfrac{q}{p}$ (C) $\dfrac{p}{q}$ (D) $-\dfrac{q}{p}$

    (E) $p + q$

20. A cube 4 inches on each side is painted red and cut into 64 1-inch cubes. How many 1-inch cubes are painted red on two faces only?

    (A) 8   (B) 12   (C) 16   (D) 24   (E) 32

21. The set $\{x / |x - L| < K\}$ is the same for all $K > 0$ and for all L, as

    (A) $\{x / 0 < x < L + K\}$ (B) $\{x / L - K < x < L + K\}$

    (C) $\{x / |L - K| < x < |L + K|\}$ (D) $\{x / |L - x| > K\}$

    (E) $\{x / -K < x < L\}$

22. Write $\left[\sqrt{2}(\cos 30° + i \sin 30°)\right]^2$ in the form $a + bi$.

    (A) $2 + i\sqrt{3}$ (B) $\dfrac{3}{2} + \dfrac{1}{2}i$ (C) $1 - i\sqrt{3}$

    (D) $\dfrac{3}{2} - \dfrac{1}{2}i$ (E) $1 + i\sqrt{3}$

23. What is the magnitude of $8 + 4i$?

    (A) 4.15   (B) 8.94   (C) 12.00   (D) 18.64   (E) 32.00

24. $\tan \dfrac{A}{2} + \cot \dfrac{A}{2}$ is equivalent to

    (A) $2 \sin A$ (B) $2 \sec A$ (C) $2 \cos A$ (D) $2 \csc A$

    (E) $2 \tan A$

25. Find the coordinates of the center of a circle whose equation is $x^2 + y^2 - 4x - 2y = 75$.

    (A) (4, 1)   (B) (1, 4)   (C) (2, 1)

    (D) (1, 2)   (E) (3, 1)

26. From two ships due east of a lighthouse and in line with its foot, the angles of elevation of the top of the lighthouse are x and y, with $x > y$. The distance between the ships is m. The distance from the lighthouse to the nearer ship is

    (A) $\dfrac{m \sin x \cos y}{\sin(x - y)}$ (B) $\dfrac{m \cos x \sin y}{\sin(x - y)}$ (C) $\dfrac{\cos x \sin y}{m \sin(x + y)}$

    (D) $m \cot x \sin y$ (E) $m \sec x \cos y$

27. What is the probability of getting 80% or more of the questions correct on a 10-question true-false exam merely by guessing?

   (A) $\dfrac{1}{16}$ (B) $\dfrac{5}{32}$ (C) $\dfrac{3}{16}$ (D) $\dfrac{7}{32}$ (E) $\dfrac{7}{128}$

28. The expression $\dfrac{3-4i}{5+3i}$ is equivalent to

   (A) $\dfrac{27-29i}{34}$ (B) $\dfrac{27-29i}{16}$ (C) $\dfrac{3-29i}{34}$ (D) $\dfrac{1}{8}$
   (E) $15 - 8i$

29. Evaluate $\lim\limits_{n \to \infty} \dfrac{3n^2}{n^2 + 10{,}000n}$.

   (A) 0   (B) 1   (C) 2   (D) 3   (E) ∞

30. If $w = w_0 e^{-kt}$, find the value of t when $w = 7$, $w_0 = 50$ and $k = 3.4$.

   (A) .52
   (B) .54
   (C) .56
   (D) .58
   (E) .60

31. Find the cube root of $27(\cos 30° + i \sin 30°)$ which, when represented graphically, lies in the second quadrant.

   (A) $3(\cos 10° + i \sin 10°)$ (B) $3(\cos 170° + i \sin 170°)$
   (C) $3(\cos 100° + i \sin 100°)$ (D) $3(\cos 130° + i \sin 130°)$
   (E) $3(\cos 150° + i \sin 150°)$

32. If $y = \dfrac{\pi}{5}$, find the value of $2 \cos \pi \sin(\pi - y) \sin\left(\dfrac{3}{2}\pi + y\right)$.

   (A) $\cos \dfrac{2}{5}\pi$ (B) $-\cos \dfrac{2}{5}\pi$ (C) $\sin \dfrac{2}{5}\pi$
   (D) $-\sin \dfrac{2}{5}\pi$ (E) $\tan \dfrac{2}{5}\pi$

DO YOUR FIGURING HERE.

33. Figure 33 is a graph of which of the following?

   (A) $x^2 + y^2 = 9$
   (B) $|x| = 3$ and $|y| = 3$
   (C) $|x + y| = 3$
   (D) $|x| + |y| = 3$
   (E) $x - y = 3$

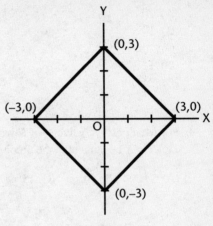

Fig. 33

34. What is the degree measure of the second quadrant angle θ for which $8 \sin^2 \theta + 6 \sin \theta = 9$?

   (A) 48.6°
   (B) 101.6°
   (C) 121.4°
   (D) 131.4°
   (E) 172.8°

35. Find the set of values satisfying the inequality $\left|\dfrac{10-x}{3}\right| < 2a$.

   (A) $4 < x < 16$ (B) $-4 > x > -16$ (C) $4 > x > -16$
   (D) $x < 16$ (E) $x > 4$

36. If the circle, $(x - 1)^2 + (y - 3)^2 = r^2$ is tangent to the line, $5x + 12y = 60$, the value of r is

   (A) $\sqrt{10}$ (B) $\dfrac{19}{13}$ (C) $\dfrac{13}{12}$ (D) $\dfrac{60}{13}$ (E) $2\sqrt{3}$

37. In a coordinate system in which the y-axis is inclined 60° to the positive x-axis, find the distance PQ between the points $P(-3, 7)$ and $Q(6, -5)$.

   (A) $\sqrt{117}$ (B) 15   (C) $\sqrt{189}$ (D) $\sqrt{333}$ (E) $\sqrt{108}$

GO ON TO THE NEXT PAGE

38. What is the remainder when $3x^4 - 2x^3 + 3x^2 - 2x + 1$ is divided by $x - 3$?

   (A) 70
   (B) 102
   (C) 200
   (D) 211
   (E) 241

39. For what positive value(s) of K will the graph of the equation $2x + y = K$ be tangent to the graph of the equation $x^2 + y^2 = 45$?

   (A) 5   (B) 10   (C) 15   (D) 20   (E) 25

40. What positive value(s) of x, less than $360°$, will give a minimum value for $4 - 2 \sin x \cos x$?

   (A) $\dfrac{\pi}{4}$ only
   (B) $\dfrac{5\pi}{4}$ only
   (C) $\dfrac{\pi}{2}$ and $\dfrac{5\pi}{2}$
   (D) $\dfrac{3\pi}{2}$
   (E) $\dfrac{\pi}{4}$ and $\dfrac{5\pi}{4}$

41. Express in radians the period of the graph of the equation $y = \dfrac{1}{3}(\cos^2 x - \sin^2 x)$.

   (A) $\dfrac{\pi}{2}$ (B) π (C) $\dfrac{3\pi}{2}$ (D) 2π (E) 3π

42. For what value of m is $4x^2 + 8xy + my^2 = 9$ the equation of a pair of straight lines?

   (A) 0   (B) 1   (C) $\dfrac{3}{2}$ (D) $\dfrac{9}{4}$ (E) 4

43. Two roots of the equation $4x^3 - px^2 + qx - 2p = 0$ are 4 and 7. What is the third root?

   (A) $\dfrac{11}{27}$ (B) $\dfrac{11}{13}$ (C) 11   (D) $\dfrac{11}{15}$ (E) $-\dfrac{22}{27}$

44. In figure 44, what is the area of parallelogram DAWN?

   (A) 11.57
   (B) 13.64
   (C) 14.63
   (D) 17.25
   (E) 20.00

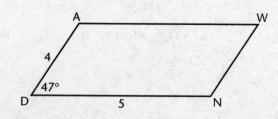

Fig. 44

45. If $\log_{6.2} x = e$, what is the value of x?

    (A) 142.54
    (B) 173.82
    (C) 227.31
    (D) 386.42
    (E) 492.75

46. If $x = 1 - e^t$ and $y = 1 + e^{-t}$, find y in terms of x.

    (A) $y = x$ (B) $y = 1 - x$ (C) $y = \dfrac{x-1}{x}$
    (D) $y = \dfrac{x}{x+1}$ (E) $y = \dfrac{2-x}{1-x}$

47. Find the value of $\log_8 \sqrt[3]{25}$.

    (A) $\dfrac{1}{2}$ (B) $\dfrac{2}{3}$ (C) $-\dfrac{2}{9}$ (D) $\dfrac{2}{9}$ (E) $-\dfrac{1}{3}$

48. If two sides of a parallelogram are 6 and 8 and one diagonal is 7, what is the length of the other diagonal?

    (A) $\sqrt{123}$ (B) $\sqrt{11}$ (C) $\sqrt{131}$ (D) $\sqrt{151}$ (E) 9

49. When $5x^{13} + 3x^{10} - K$ is divided by $x + 1$, the remainder is 20. The value of K is

    (A) -22 (B) -12 (C) 8 (D) 28 (E) 14

50. What is the smallest possible value of x (in degrees) for which $\cos x - \sin x = \dfrac{1}{\sqrt{2}}$?

    (A) 5° (B) 12° (C) 15° (D) 18° (E) 30°

STOP

Sample Test 1
Answer Key

Math Level IIC

1. B	11. B	21. B	31. D	41. B
2. E	12. E	22. E	32. C	42. E
3. E	13. E	23. B	33. D	43. B
4. D	14. E	24. D	34. D	44. C
5. C	15. C	25. C	35. A	45. A
6. B	16. E	26. B	36. B	46. E
7. B	17. A	27. E	37. A	47. C
8. D	18. B	28. C	38. D	48. D
9. C	19. A	29. D	39. C	49. A
10. A	20. D	30. D	40. E	50. C

Solutions

1. **(B)** $\sqrt{x-3} = x - 9$ Square both sides
$$x - 3 = x^2 - 18x + 81$$
$$x^2 - 19x + 84 = 0$$
$$(x - 12)(x - 7) = 0$$
$$x = 12 \text{ and } x = 7$$

Check: $\sqrt{12-3} = 12 - 9$
$\sqrt{9} = 3$
only 1 root

Check: $\sqrt{4} = -2$
does not check
reject $x = 7$

2. **(E)** $\left(\frac{1}{2}\right)^3 \square (3)^{\frac{1}{2}}$

$= \frac{1}{8} \square \sqrt{3}$

$= \left(\frac{1}{8}\right)^{\sqrt{3}} - \left(\sqrt{3}\right)^{\frac{1}{8}}$

$= .027277 - 1.07107$

$= -1.044$

3. **(E)** $f(-2x) = 3(-2x)^2\ 5(-2x) - 4$
$= 12x^2 + 10x - 4$

This is no multiple of the original function.

4. **(D)** $\dfrac{P}{K} = e^{-st}$ or $-tx = \log_e \dfrac{P}{K}$

$$x = -\dfrac{1}{t}\left(\log_e P - \log_e K\right) = \dfrac{\log_e K - \log_e P}{t}$$

$$x = \dfrac{\log K - \log P}{t \log e}$$

5. **(C)** Slope of $x = 3y$ is $1/3$.

   Slope of $y = -3x + 7$ is -3. Hence, since slopes are negative reciprocals, the lines are $\perp$, and the $\triangle$ is right.

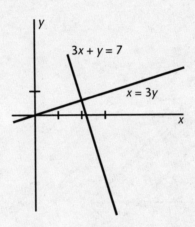

6. **(B)** $x^2 - 6x + 9 + y^2 + 8y + 16 = 25$

   $(x-3)^2 + (y+4)^2 = 25$

   Curve is circle of radius 5.
   Hence, area is 25π.

7. **(B)** Let each side of cube $= 1$

   then diagonal of cube: $D = \sqrt{1^2 + 1^2 + 1^2}$
   $= \sqrt{3}$

   diagonal of face: $D' = \sqrt{1^2 + 1^2} = \sqrt{2}$

   $\dfrac{D}{D'} = \dfrac{\sqrt{3}}{\sqrt{2}} \cdot \dfrac{\sqrt{3}}{\sqrt{3}} = \dfrac{3}{\sqrt{6}}$

8. **(D)** $\sin\left(22\frac{1}{2}\right)^\circ = \frac{x}{2} = \sin\frac{45^\circ}{2}$

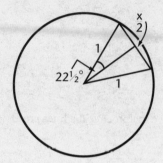

$$\frac{\sqrt{1-\cos 45^\circ}}{2} = \frac{x}{2}$$

$$\frac{\sqrt{1-\frac{\sqrt{2}}{2}}}{2} = \frac{\frac{\sqrt{2}-1}{\sqrt{2}}}{2} = \frac{x}{2}$$

$$x = \sqrt{2-\sqrt{2}}$$

9. **(C)**

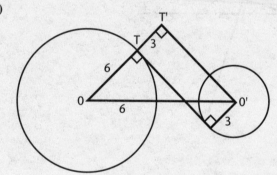

Extend OT 3" to T' and draw $O'T'$. Then in right $\triangle OT'O'$, $OO' = 15"$ and $OT' = 9"$ so that $O'T' = 12"$.

10. **(A)** The cosine function $y = \cos x$ has a period of 2π radians. Hence $y = 5\cos 3x$ has a period of $\frac{2\pi}{3}$ radians.

11. **(B)**
$2x = 2^{3(y+1)}$ $\quad$ $3^{2y} = 3^{x-9}$
$x = 3y + 3$ $\quad\quad$ $2y = x - 9$
$\quad\quad\quad\quad\quad\quad$ $2y = 3y + 3 - 9$
$\quad\quad\quad\quad\quad\quad$ $y = 6$

12. (E)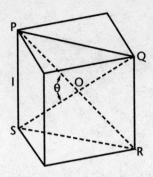

Let each edge = 1

PQRS is a rectangle

$PR = SQ = \sqrt{3}$

$PO = SO = \dfrac{\sqrt{3}}{2}$

then $1^2 = \left(\dfrac{\sqrt{3}}{2}\right)^2 + \left(\dfrac{\sqrt{3}}{2}\right)^2 - 2\dfrac{\sqrt{3}}{2}\dfrac{\sqrt{3}}{2}\cos\theta$

$\dfrac{3}{2}\cos\theta = 1/2$

$\cos\theta = 1/3$

$\theta = \cos^{-1} 1/3$

13. (E) $x^2 y = 10^{\log x}$

$\log x^2 y = \log x$

$x^2 y = x$

$xy = 1$

$y = \dfrac{1}{x}$

y varies inversely with x.

14. (E) $\log_r\left(\dfrac{r}{2}\right) = \log_r r - \log_r 2 = 1 - \log_r 2$

$m - n = \log_r 6 - \log_r 3 = \log_r \dfrac{6}{3} = \log_r 2$

$\log_r\left(\dfrac{r}{2}\right) = 1 - (m - n) = 1 - m + n$

15. **(C)** $\quad x^2 + x - 6 < 0$
$(x+3)(x-2) < 0$
Either $(x+3) > 0$ and $(x-2) < 0$
or $(x+3) < 0$ and $(x-2) > 0$
Either $x > -3$ and $x < 2$ $\qquad -3 < x < 2$
or $x < -3$ and $x > 2$ $\qquad$ Impossible
Therefore, $-3 < x < 2$.

16. **(E)** The contrapositive is the converse of the inverse. Thus, form the converse and negate the hypothesis and conclusion, Hence $\sim q \to p$.

17. **(A)** $\quad \sqrt{x^2 + y^2} = 2\sqrt{(x-3)^2 + y^2}$
$x^2 + y^2 = 4(x^2 - 6x + 9 + y^2)$
$3x^2 - 24x + 36 + 3y^2 = 0$
$x^2 - 8x + y^2 = -12 \quad$ a circle

18. **(B)** Solve for x in terms of y:
$xy - 3y = 2x + 1$
$xy - 2x = 3y + 1$
$x = \dfrac{3y+1}{x-2}$

Now interchange x and y.
$y = \dfrac{3x+1}{x-2}$

Hence $K = 2$.

19. **(A)** Let the roots be r and s.
Then $r + s = -p$ and $rs = q$.
$\dfrac{1}{r} + \dfrac{1}{s} = \dfrac{r+s}{rs} = -\dfrac{p}{q}$

20. **(D)** A 1-inch cube will be painted on two sides only if it lies on one edge of the 4-inch cube, but does not touch a vertex of the original cube. On each edge there are two such cubes. Since a cube has 12 edges, there are 24 such cubes.

21. **(B)** If $x > L$, then $|x - L| < K$ means $x - L < K$ or $x < L + K$
    If $x < L$, then $|x - L| < K$ means $L - x < K$
    or $-x < K - L$ or $x > L - K$
    so that $L - K < x < L + K$

22. **(E)** By De Moivre's Theorem,
    $[\sqrt{2}(\cos 30° + i \sin 30°)]^2 = 2(\cos 60° + i \sin 60°)$
    $$= 2\left(\frac{1}{2} + i\frac{\sqrt{3}}{2}\right)$$
    $$= 1 + i\sqrt{3}$$

23. **(B)** The magnitude $= |8 + 4i| = \sqrt{8^2 + 4^2}$
    $$= \sqrt{80}$$
    $$= 8.94$$

24. **(D)** $\tan\frac{A}{2} + \cot\frac{A}{2} = \tan A + \dfrac{1}{\tan\frac{A}{2}}$

    $$= \frac{\tan^2\frac{A}{2} + 1}{\tan\frac{A}{2}} = \frac{\sec^2\frac{A}{2}}{\tan\frac{A}{2}}$$

    $$= \frac{1}{\cos^2\frac{A}{2}} \cdot \frac{\cos\frac{A}{2}}{\sin\frac{A}{2}}$$

    $$= \frac{1}{\sin\frac{A}{2}\cos\frac{A}{2}}$$

    $$= \frac{1}{\frac{1}{2}\sin A} = \frac{2}{\sin A}$$

    $$= 2\csc A$$

25. **(C)** $x^2 - 4x + 4 + y^2 - 2y + 1 = 75 + 4 + 1$
    $(x - 2)^2 + (y - 1)^2 = 80$

    center is at $(2, 1)$.

26. **(B)** In $\triangle PQR$, by law of sines,

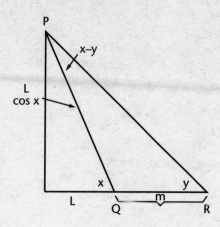

$$\frac{m}{\sin(x-y)} = \frac{\frac{L}{\cos x}}{\sin y}$$

$$m \sin y = \frac{L \sin(x-y)}{\cos x}$$

$$\frac{m \cos x \sin y}{\sin(x-y)} = L$$

27. **(E)** May get 8 or 9 or 10 correct

$$\text{Probability of getting 10 right} = \left(\frac{1}{2}\right)^{10}$$

Probability of getting 9 right =

$$^{10}C_1 \left(\frac{1}{2}\right)^9 \left(\frac{1}{2}\right) =$$

$$10\left(\frac{1}{2}\right)^{10}$$

Probability of getting 8 right =

$$^{10}C_2 \left(\frac{1}{2}\right)^8 \left(\frac{1}{2}\right)^2 = \frac{10}{1} \cdot \frac{9}{2}\left(\frac{1}{2}\right)^{10} =$$

$$45\left(\frac{1}{2}\right)^{10}$$

Probability of getting 8 or 9 or 10 right =

$$\left(\frac{1}{2}\right)^{10} + 10\left(\frac{1}{2}\right)^{10} + 45\left(\frac{1}{2}\right)^{10} =$$

$$\left(\frac{1}{2}\right)^{10} \cdot 56 = \frac{7}{2^7} = \frac{7}{128}$$

28. **(C)** $\dfrac{3-4i}{5+3i} \cdot \dfrac{5-3i}{5-3i} = \dfrac{15-29i-12}{25+9} = \dfrac{3-29i}{34}$

29. **(D)** Divide numerator and denominator by n^2.

$$\lim_{n \to \infty} \dfrac{3n^2}{n^2 + 10,000n} = \lim_{n \to \infty} \dfrac{3}{1 + \dfrac{10,000}{n}} = \dfrac{3}{1} = 3$$

30. **(D)**

$$w = w_0 e^{-kt}$$
$$7 = 50 e^{-3.4t}$$
$$\dfrac{7}{50} = e^{-3.4t}$$
$$\ln\left(\dfrac{7}{50}\right) = \ln e^{-3.4t} = -3.4t$$
$$\Downarrow$$
$$-3.4t = \ln\left(\dfrac{7}{50}\right)$$
$$t = \dfrac{\ln\left(\dfrac{7}{50}\right)}{-3.4} = \dfrac{-1.966}{-3.4} = .578$$

31. **(D)**

$27 (\cos 30° + i \sin 30°) = 27 (\cos 390° + i \sin 390°)$
$[27 (390° + i \sin 390°)]^{1/3} = 3 (\cos 130° + i \sin 130°)$

32. **(C)**

$\cos \pi = -1$, $\sin (\pi - y) = \sin y$

$\sin\left(\dfrac{3}{2}\pi + y\right)$

$\quad = \sin\dfrac{3}{2}\pi \cos y + \cos\dfrac{3}{2}\pi \sin y$

$\quad = (-1)\cos y + 0 = -\cos y$

$\quad 2 \cos \pi \sin (\pi - y) \sin\left(\dfrac{3}{2}\pi + y\right)$

$\quad = 2(-1) \sin y (-\cos y)$

$\quad = 2 \sin y \cos y = \sin 2y$

$\quad = \sin \dfrac{2\pi}{5}$

33. **(D)** (A) graphs as a circle;
(B) graphs as vertical and horizontal lines;
(C) $|x+y|=3$ consists of 2 lines $x+y=3$ and $-x-y=3$;
(E) graphs as one straight line; and
(D) graphs as $x+y=3$, $x-y=3$, $-x+y=3$,
and $x+y=3$, which are the four lines in the graph.

34. **(D)**
$$8\sin^2\theta + 6\sin\theta - 9 = 0$$
$$(4\sin\theta - 3)(2\sin\theta + 3) = 0$$

$\sin\theta = \dfrac{3}{4}$ $\quad\bigg|\quad$ $\sin\theta = -\dfrac{3}{2}$

$\theta = \sin^{-1}\dfrac{3}{4}$ $\quad\bigg|\quad$ reject

The second quadrant
solution is $180° - 48.6°$
$= 131.4°$

35. **(A)** $\left|\dfrac{10-x}{3}\right| < 2$

$|10 - x| < 6$

$-6 < 10 - x < 6$

$-16 < -x < -4$

or $4 < x < 16$

36. **(B)** The center of the circle is (1, 3). The value of r is then equal to the distance from the center to the given line. Thus

$$r = \left|\dfrac{5x_1 + 12y_1 - 60}{\sqrt{5^2 + 12^2}}\right| = \left|\dfrac{5(1) + 12(3) - 60}{13}\right|$$

$$r = \dfrac{19}{13}$$

37. **(A)**

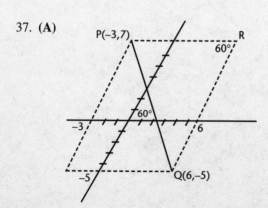

From the figure, PQ is the side of $\triangle PQR$ opposite $\angle R = 60°$.
$PR = 9$ and $QR = 12$. Thus
$PQ^2 = 9^2 + 12^2 - 2 \cdot 9 \cdot 12 \cos 60°$
$= 225 - 108$
$= 117$
$PQ = \sqrt{117}$

38. **(D)** $3(3)^4 - 2(3)^3 + 3(3)^2 - 2(3) + 1 = 211$

or

$$\begin{array}{rrrrr|r} 3 & -2 & 3 & -2 & 1 & \underline{3} \\ & 9 & 21 & 72 & 210 & \\ \hline 3 & 7 & 24 & 70 & \underline{211} \end{array}$$

or

$$x - 3 \overline{\smash{\big)}\, 3x^4 - 2x^3 + 3x^2 - 2x + 1} \quad 3x^3 + 7x^2 + 24x + 70 + \frac{211}{x-3}$$

39. **(C)**
$$y = K - 2x$$
$$x^2 + (K - 2x)^2 = 45$$
$$x^2 + 4x^2 - 4Kx + (K^2 - 45) = 0$$
$$5x^2 - 4Kx + (K^2 - 45) = 0$$

If the line is tangent, the quadratic equation will have two equal roots. Thus the discriminant = 0.

$$16K^2 - 20(K^2 - 45) = 0$$
$$4K^2 = 900$$
$$2K = 30$$
$$K = 15$$

40. **(E)** If $y = 4 - 2 \sin x \cos x = 4 - \sin 2x$, y will be a minimum when $\sin 2x$ is at a maximum; that is, at

$$2x = \frac{\pi}{2} \text{ and } \frac{5\pi}{2}$$
$$\text{or } x = \frac{\pi}{4} \text{ and } \frac{5\pi}{4}$$

41. **(B)** $y = \frac{1}{3}(\cos^2 x - \sin^2 x) = \frac{1}{3}\cos 2x$

Since $\cos x$ has a period of 2π radians, $\cos 2x$ has a period of π.

42. **(E)** In order to make the left member a perfect square, m must equal 4. Then

$$4x^2 + 8xy + 4y^2 = 4(x+y)^2 = 9$$

or

$$(x+y)^2 = \frac{9}{4}$$

and

$$x + y = \pm\frac{3}{2}$$

which graphs as a pair of straight lines.
Thus $m = 4$.

43. **(B)**   Let r be the root, then

$$4+7+r=\frac{p}{4}=11+r$$

$$4\cdot 7\cdot r=\frac{p}{2}=28r \quad \text{or} \quad \frac{p}{4}=14r$$

Thus $14r = 11 + r$ and $r = \frac{11}{13}$

44. **(C)**
$$A = ab \sin C$$
$$= \left(\overline{AD}\right)\left(\overline{DN}\right)\sin D$$
$$= (4)(5) \sin 47°$$
$$= 20 \sin 47°$$
$$= 14.627$$

45. **(A)**
$$\log_{6.2} x = e$$
$$\Downarrow$$
$$x = (6.2)^e$$
recall $e = 2.71828...$
$$x = 142.54$$

46. **(E)** $y = 1 + \dfrac{1}{e^t}$ and $e^t = 1 - x$

$$y = 1 + \frac{1}{1-x} = \frac{1-x+1}{1-x} = \frac{2-x}{1-x}$$

47. **(C)**   Let $x = \log_8 \sqrt[3]{.25}$

then $8^x = \sqrt[3]{.25} = \dfrac{1}{4^{1/3}}$

or $2^{3x} = 2^{-2/3}$

$$3x = -\frac{2}{3}$$

$$x = -\frac{2}{9}$$

48. **(D)**

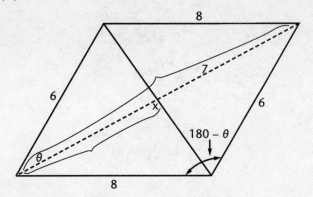

$$7^2 = 6^2 + 8^2 - 2 \cdot 6 \cdot 8 \cos \theta$$
$$49 = 100 - 96 \cos \theta$$
$$\cos \theta = \frac{15}{96}$$
$$x^2 = 6^2 + 8^2 - 2 \cdot 6 \cdot 8 \cos(180 - \theta)$$
$$= 100 - 96 \cos(180 - \theta)$$
$$x^2 = 100 + 96 \cos \theta = 100 + 96\left(\frac{51}{96}\right)$$
$$x = \sqrt{151}$$

49. **(A)** $P(x) = 5x^{13} + 3x^{10} - K$
$P(-1) = 5(-1)^{13} + 3(-1)^{10} - K = 20$
$-5 + 3 - K = 20$
$K = -22$

50. **(C)** Square both sides of the equation.

$$(\cos x - \sin x)^2 = \frac{1}{2}$$
$$\cos^2 x + \sin^2 x - 2 \sin x \cos x = \frac{1}{2}$$
$$1 - \frac{1}{2} = \sin 2x = \frac{1}{2}$$
$$2x = 30°$$
$$x = 15°$$

Sample Test 2
Answer Sheet

Math Level IIC

1. Ⓐ Ⓑ Ⓒ Ⓓ Ⓔ
2. Ⓐ Ⓑ Ⓒ Ⓓ Ⓔ
3. Ⓐ Ⓑ Ⓒ Ⓓ Ⓔ
4. Ⓐ Ⓑ Ⓒ Ⓓ Ⓔ
5. Ⓐ Ⓑ Ⓒ Ⓓ Ⓔ
6. Ⓐ Ⓑ Ⓒ Ⓓ Ⓔ
7. Ⓐ Ⓑ Ⓒ Ⓓ Ⓔ
8. Ⓐ Ⓑ Ⓒ Ⓓ Ⓔ
9. Ⓐ Ⓑ Ⓒ Ⓓ Ⓔ
10. Ⓐ Ⓑ Ⓒ Ⓓ Ⓔ
11. Ⓐ Ⓑ Ⓒ Ⓓ Ⓔ
12. Ⓐ Ⓑ Ⓒ Ⓓ Ⓔ
13. Ⓐ Ⓑ Ⓒ Ⓓ Ⓔ
14. Ⓐ Ⓑ Ⓒ Ⓓ Ⓔ
15. Ⓐ Ⓑ Ⓒ Ⓓ Ⓔ
16. Ⓐ Ⓑ Ⓒ Ⓓ Ⓔ
17. Ⓐ Ⓑ Ⓒ Ⓓ Ⓔ
18. Ⓐ Ⓑ Ⓒ Ⓓ Ⓔ
19. Ⓐ Ⓑ Ⓒ Ⓓ Ⓔ
20. Ⓐ Ⓑ Ⓒ Ⓓ Ⓔ
21. Ⓐ Ⓑ Ⓒ Ⓓ Ⓔ
22. Ⓐ Ⓑ Ⓒ Ⓓ Ⓔ
23. Ⓐ Ⓑ Ⓒ Ⓓ Ⓔ
24. Ⓐ Ⓑ Ⓒ Ⓓ Ⓔ
25. Ⓐ Ⓑ Ⓒ Ⓓ Ⓔ
26. Ⓐ Ⓑ Ⓒ Ⓓ Ⓔ
27. Ⓐ Ⓑ Ⓒ Ⓓ Ⓔ
28. Ⓐ Ⓑ Ⓒ Ⓓ Ⓔ
29. Ⓐ Ⓑ Ⓒ Ⓓ Ⓔ
30. Ⓐ Ⓑ Ⓒ Ⓓ Ⓔ
31. Ⓐ Ⓑ Ⓒ Ⓓ Ⓔ
32. Ⓐ Ⓑ Ⓒ Ⓓ Ⓔ
33. Ⓐ Ⓑ Ⓒ Ⓓ Ⓔ
34. Ⓐ Ⓑ Ⓒ Ⓓ Ⓔ
35. Ⓐ Ⓑ Ⓒ Ⓓ Ⓔ
36. Ⓐ Ⓑ Ⓒ Ⓓ Ⓔ
37. Ⓐ Ⓑ Ⓒ Ⓓ Ⓔ
38. Ⓐ Ⓑ Ⓒ Ⓓ Ⓔ
39. Ⓐ Ⓑ Ⓒ Ⓓ Ⓔ
40. Ⓐ Ⓑ Ⓒ Ⓓ Ⓔ
41. Ⓐ Ⓑ Ⓒ Ⓓ Ⓔ
42. Ⓐ Ⓑ Ⓒ Ⓓ Ⓔ
43. Ⓐ Ⓑ Ⓒ Ⓓ Ⓔ
44. Ⓐ Ⓑ Ⓒ Ⓓ Ⓔ
45. Ⓐ Ⓑ Ⓒ Ⓓ Ⓔ
46. Ⓐ Ⓑ Ⓒ Ⓓ Ⓔ
47. Ⓐ Ⓑ Ⓒ Ⓓ Ⓔ
48. Ⓐ Ⓑ Ⓒ Ⓓ Ⓔ
49. Ⓐ Ⓑ Ⓒ Ⓓ Ⓔ
50. Ⓐ Ⓑ Ⓒ Ⓓ Ⓔ

Directions: For each question in the sample test, select the best of the answer choices and blacken the corresponding space on this answer sheet.

Please note: (a) You will need to use a calculator in order to answer some, though not all, of the questions in this test. As you look at each question, you must decide whether or not you need a calculator for the specific question. A four-function calculator is not sufficient; your calculator must be at least a scientific calculator. Calculators that can display graphs and programmable calculators are also permitted.

(b) Set your calculator to radian mode or degree mode depending on the requirements of the question.

(c) All figures are accurately drawn and are intended to supply useful information for solving the problems that they accompany. Figures are drawn to scale UNLESS it is specifically stated that a figure is not drawn to scale. Unless otherwise indicated, all figures lie in a plane.

(d) The domain of any function f is assumed to be the set of all real numbers x for which $f(x)$ is a real number except when this is specified not to be the case.

(e) Use the reference data below as needed.

REFERENCE DATA

SOLID	VOLUME	OTHER		
Right circular cone	$V = \frac{1}{3}\pi r^2 h$	$S = cl$	V = volume r = radius h = height	S = lateral area c = circumference of base l = slant height
Sphere	$V = \frac{4}{3}\pi r^3$	$S = 4\pi r^2$	V = volume r = radius S = surface area	
Pyramid	$V = \frac{1}{3}Bh$		V = volume B = area of base h = height	

TEAR HERE

Sample Test 2

MATH LEVEL IIC

50 Questions • Time—60 Minutes

1. The fraction $\frac{1}{1+i}$ is equivalent to

   (A) $1-i$ (B) $\frac{1+i}{2}$ (C) $\frac{1-i}{2}$ (D) i
   (E) $-i$

2. Find the value of the remainder obtained when $6x^4 + 5x^3 - 2x + 8$ is divided by $x - \frac{1}{2}$.

   (A) 2   (B) 4   (C) 6   (D) 8   (E) 10

3. Solve the equation $2x + \sqrt{x} - 1 = 0$.

   (A) 1 and 1/4   (B) 1/4   (C) 1   (D) 0   (E) 4

4. Solve for r: $27^{6-r} = 9^{r-1}$

   (A) 1   (B) 2   (C) 3   (D) 4   (E) 5

5. How many integers greater than 1000 can be formed from the digits 0, 2, 3, 5, if no digit is repeated in any number?

   (A) 9   (B) 18   (C) 27   (D) 36   (E) 72

6. What is $\lim\limits_{x \to \infty} \frac{3x^2 - \sqrt{5}x + 4}{2x^2 + 3x + \sqrt{11}}$?

   (A) -1.12
   (B) .91
   (C) 1.11
   (D) 1.33
   (E) 1.50

7. Find the radius of the circle whose equation is
$$x^2 + y^2 - 6x + 8y = 0$$
   (A) 1   (B) 2   (C) 3   (D) 4   (E) 5

8. When drawn on the same set of axes, the graphs of $x^2 - 3y^2 = 9$ and $(x - 2)^2 + y^2 = 9$ have in common exactly
   (A) 0 points
   (B) 1 point
   (C) 2 points
   (D) 3 points
   (E) 4 points

9. If the equation $x^3 - 6x^2 + px + q = 0$ has 3 equal roots, then
   (A) $q = 0$
   (B) $p = 0$
   (C) $q = 2$
   (D) each root = 2
   (E) each root = –2

10. A root of $x^5 - 32 = 0$ lies in quadrant II. Write this root in polar form.
    (A) $2(\cos 120° + i \sin 120°)$
    (B) $2(\cos 144° + i \sin 144°)$
    (C) $2(\cos 150° + i \sin 150°)$
    (D) $4(\cos 144° + i \sin 144°)$
    (E) $2(\cos 72° + i \sin 72°)$

11. The solution set of $x^2 < 3x + 10$ is given by the inequality
    (A) $x < 5$
    (B) $x > -2$
    (C) $-2 < x < 5$
    (D) $-2 \leq x \leq 5$
    (E) $x > 5$

12. Write the complete number $-2 - 2i$ in polar form.
    (A) $2\left(\cos \dfrac{\pi}{4} + i \sin \dfrac{\pi}{4}\right)$
    (B) $-2\left(\cos \dfrac{\pi}{4} - i \sin \dfrac{\pi}{4}\right)$
    (C) $2\sqrt{2}\left(\cos \dfrac{3\pi}{4} + i \sin \dfrac{3\pi}{4}\right)$
    (D) $2\sqrt{2}\left(\cos \dfrac{7\pi}{4} + i \sin \dfrac{7\pi}{4}\right)$
    (E) $2\sqrt{2}\left(\cos \dfrac{5\pi}{4} + i \sin \dfrac{5\pi}{4}\right)$

13. If $0 < x < 1$, then
    (A) $0 < \log_{10} x < 1$
    (B) $\log_{10} x > 1$
    (C) $\log_{10} x < 0$
    (D) $\log_{10} x < -1$
    (E) none of these is true

14. The inverse of $\sim p \rightarrow \sim q$ is equivalent to
    (A) $p \rightarrow \sim q$
    (B) $q \rightarrow p$
    (C) $q \rightarrow \sim p$
    (D) $p \rightarrow q$
    (E) $\sim q \rightarrow p$

DO YOUR FIGURING HERE.

15. If $f(x,y) = (\ln x^2)(e^{2y})$, what is the value of $f(2, \sqrt{2})$?

    (A) 23.45
    (B) 24.35
    (C) 25.34
    (D) 25.43
    (E) 27.25

16. If x and y are elements in the set of real numbers, which is *not* a function?

    (A) $f = \{(x,y)/y = x^2 + 1\}$
    (B) $f = \{(x,y)/y = 2x^3\}$
    (C) $f = \{(x,y)/y = 9 - x^2\}$
    (D) $f = \{(x,y)/y \geq x + 1\}$
    (E) $f = \{(x,y)/y = x^2 - |x|\}$

17. If $\frac{1}{x} + y = 2$ and $x + \frac{1}{y} = 3$, then the ratio of x to y is

    (A) 1:2   (B) 2:3   (C) 3:1   (D) 3:2   (E) 3:4

18. If $f(x) = 3x^2 - 2x + 5$, find $\lim\limits_{h \to 0} \frac{f(x+h) - f(x)}{h}$

    (A) $6x - 2$ (B) 0
    (C) ∞ (D) indeterminate
              (E) 5

19. Evaluate $\log_{11} 21$

    (A) 1.27
    (B) 1.21
    (C) 1.18
    (D) 1.15
    (E) 1.02

20. The focus of a parabola is the point (0, 2) and its directrix is the line $y = -2$. Write an equation of the parabola.

    (A) $y^2 = 8x$ (B) $x^2 = 8y$
    (C) $x^2 = 4y$ (D) $y^2 = 4x$
              (E) $x^2 = 2y$

21. Find the positive value of $\sin(\tan^{-1} 3)$.

    (A) $\frac{3}{10}$ (B) $\frac{3}{5}$ (C) $\frac{3}{\sqrt{10}}$ (D) $\frac{1}{3}$ (E) $\frac{1}{\sqrt{10}}$

22. As angle x increases from 0 to 2π radians, tan x increases in

(A) no quadrants
(B) the first and third quadrants only
(C) the second and fourth quadrants only
(D) all four quadrants
(E) the first and second quadrants only

23. A pyramid is cut by a plane parallel to its base at a distance from the base equal to two-thirds the length of the altitude. The area of the base is 18. Find the area of the section determined by the pyramid and the cutting plane.

(A) 1   (B) 2   (C) 3   (D) 6   (E) 9

24. If $\sqrt{23} \sin x = \sqrt{17}$ and $\cos x > 0$, what is the value of $\tan x$?

(A) 1.28
(B) 1.35
(C) 1.68
(D) 1.79
(E) 2.03

25. The point whose polar coordinates are $(5, -30°)$ is the same as the point whose polar coordinates are

(A) $(-5, 30°)$ (B) $(-5, 150°)$
(C) $(5, -150°)$ (D) $(-5, 30°)$
         (E) $(5, 150°)$

26. A coin is tossed three times. Find the probability of the event represented by the composite statement $\sim p \wedge q$ if

   p: exactly two heads show
   q: at least two heads show

(A) $\frac{1}{2}$ (B) $\frac{1}{4}$ (C) $\frac{1}{6}$ (D) $\frac{1}{8}$ (E) $\frac{3}{4}$

27. A rod, pivoted at one end, rotates through $\frac{2\pi}{3}$ radians. If the rod is 6 inches long, how many inches does the free end travel?

(A) π (B) 2π (C) 3π (D) 4π (E) $\frac{3\pi}{2}$

28. A value that satisfies the equation $\cos^2 x - 2\cos x = 0$ is (in degrees)

    (A) 0   (B) 30   (C) 60   (D) 90   (E) none of these

29. $\cot\theta - \dfrac{\cos 2\theta}{\sin\theta \cos\theta} =$

    (A) $\sin\theta$ (B) $\cos\theta$
    (C) $\tan\theta$ (D) $\cot\theta$
             (E) $\sec\theta$

30. If $m > 1$, the maximum value of $2m \sin 2x$ is

    (A) 2   (B) m
    (C) $2m$ (D) $4m$
          (E) none of these

31. If $x(t) = 3\cos t$
    $y(t) = 2 + 4\sin t$, what is the value of x when $y = 5$?

    (A) 1.98
    (B) 1.78
    (C) 1.58
    (D) 1.38
    (E) 1.18

32. The graph of $y = x - |x|$ is equivalent to the graph of

    (A) $y = x$
    (B) $y = 2x$
    (C) $y = 2x$ for $0 \leq x \leq 1$
    (D) $y = 2x$ for $x \leq 0$
    (E) $y = 2x$ for $x \geq 0$

33. The function of $f(x) = \dfrac{1}{\sqrt{x+1}}$ is defined for $\{x | x$ is a real number and $x > -1\}$.

    Write an expression for the inverse of $f(x)$.

    (A) $\dfrac{1-x^2}{x^2}$ such that $x > 0$

    (B) $\dfrac{1-x}{x}$

    (C) $\dfrac{x^2-1}{x^2}$

    (D) $\dfrac{1-x^2}{x^2}$ such that $x > -1$

    (E) none of these

DO YOUR FIGURING HERE.

34. One side of a given triangle is 18 inches. Inside the triangle a line segment is drawn parallel to this side cutting off a triangle whose area is two-thirds of that of the given triangle. Find the length of this segment in inches.

    (A) 12 (B) $6\sqrt{6}$ (C) $9\sqrt{2}$ (D) $6\sqrt{3}$ (E) 9

35. What is the x-intercept of $f(x) = \dfrac{x+\sqrt{17}}{x-1.6}$?

    (A) −2.35
    (B) −2.65
    (C) −2.95
    (D) −3.25
    (E) −3.55

36. How many real roots does the following equation have?
    $$e^x - e^{-x} + 1 = 0$$

    (A) 0 (B) 1 (C) 2 (D) 4 (E) an infinite number

37. The graph of $|x| + |y|$ consists of

    (A) one straight line
    (B) a pair of straight lines
    (C) the sides of a square
    (D) a circle
    (E) a point

38. If the perimeter of an isosceles triangle is 36 and the altitude to the base is 6, find the length of the altitude to one of the legs.

    (A) 4.8
    (B) 6
    (C) 9.6
    (D) 10
    (E) Cannot be found on the basis of the given data

39. If the radius of a sphere is doubled, the percent increase in volume is

    (A) 100 (B) 200 (C) 400 (D) 700 (E) 800

40. Two different integers are selected at random from the integers 1 to 12 inclusive. What is the probability that the sum of the two numbers is even?

    (A) $\dfrac{1}{18}$ (B) $\dfrac{1}{2}$ (C) $\dfrac{5}{9}$ (D) $\dfrac{4}{11}$ (E) $\dfrac{5}{11}$

41. The equality $\frac{3}{\sec^2 x} = 5 - \frac{3}{\csc^2 x}$ is satisfied by

    (A) all values of x
    (B) exactly two values of x
    (C) only one value of x
    (D) no value of x
    (E) infinitely many but not all values of x

42. If the first term of a geometric progression is $\sqrt[3]{m}$ and the third term is $\sqrt{m}$, what is the 13th term of the progression?

    (A) m (B) $2m$ (C) $m^{4/3}$ (D) $m^{5/3}$ (E) m^2

43. Lines AB and AC are tangents to a circle at points B and C respectively. Minor arc BC is 7π inches, and the radius of the circle is 18 inches. What is the number of degrees in angle BAC?

    (A) 90°   (B) 95°   (C) 70°   (D) 100°   (E) 110°

44. Two spheres, of radius 8 and 2, are resting on a plane table top so that they touch each other.

    How far apart are their points of contact with the plane table top?

    (A) 6   (B) 7   (C) 8   (D) $8\sqrt{2}$ (E) 9

45. The hyperbola $\frac{y^2}{11} - \frac{x^2}{9} = 1$ intersects the y-axis at which of the following points?

    (A) (0, 4.23)
    (B) (0, 3.84)
    (C) (0, 3.32)
    (D) (0, 1.32)
    (E) (−3.32, 0)

46. If n is an integer, what is the remainder when
    $5x^{2n+1} - 10x^{2n} + 3x^{2n-1} + 5$
    is divided by $x + 1$?

    (A) 0   (B) 2   (C) 4   (D) −8   (E) −13

DO YOUR FIGURING HERE.

47. If the roots of the equation $x^2 - px + q = 0$ are r_1 and r_2, then $r_1^2 + r_2^2 =$

   (A) $p^2 + q^2$
   (B) $p^2 - 2q$
   (C) $p^2 - q^2$
   (D) p^2
   (E) q^2

48. Find the value, in simplest form, of

$$\frac{2^{n+4} - 2(2^n)}{2(2^{n+3})}$$

   (A) $\frac{1}{2}$ (B) $\frac{1}{4}$ (C) $\frac{3}{4}$ (D) $\frac{5}{8}$ (E) $\frac{7}{8}$

49. Find the area of the regular octagon inscribed in a circle of radius 8.

   (A) 18 (B) $26\sqrt{2}$ (C) 120 (D) $128\sqrt{2}$ (E) $100\sqrt{3}$

50. What is the number of radians of the smallest positive angle x which will give the maximum value for $y = 3 - \cos 2x$?

   (A) $\frac{\pi}{4}$ (B) $\frac{\pi}{2}$ (C) π (D) $\frac{3\pi}{2}$ (E) 2π

DO YOUR FIGURING HERE.

STOP

Sample Test 2
Answer Key

Math Level IIC

1. C	11. C	21. C	31. A	41. D
2. D	12. E	22. D	32. D	42. C
3. B	13. C	23. B	33. A	43. E
4. D	14. D	24. C	34. B	44. C
5. B	15. A	25. B	35. B	45. C
6. E	16. D	26. D	36. B	46. E
7. E	17. D	27. D	37. C	47. B
8. C	18. A	28. D	38. C	48. E
9. D	19. A	29. C	39. D	49. D
10. B	20. B	30. C	40. E	50. B

Solutions

1. **(C)** $\dfrac{1}{1+i} \cdot \dfrac{1-i}{1-i} = \dfrac{1-i}{1-i^2} = \dfrac{1-i}{2}$

2. **(D)** Remainder $= 6\left(\dfrac{1}{2}\right)^4 + 5\left(\dfrac{1}{2}\right)^3 - 2\left(\dfrac{1}{2}\right) + 8$

    $= 6 \cdot \dfrac{1}{16} + 5 \cdot \dfrac{1}{8} - 1 + 8$

    $= \dfrac{3}{8} + \dfrac{5}{8} - 1 + 8 = 8$

3. **(B)** $\sqrt{x} = 1 - 2x$

    Square both sides:
    $$x = 1 - 4x + 4x^2$$
    $$4x^2 - 5x + 1 = 0$$
    $$(4x - 1)(x - 1) = 0$$
    $$x = 1/4, \qquad x = 1$$

    Check $x = 1/4$, $\sqrt{1/4} = 1 - 2(1/4)$
    $$\dfrac{1}{2} = \dfrac{1}{2}$$

    Check $x = 1$, $\sqrt{1} = 1 - 2(1)$ or $1 = -1$

    (does not check)

    Thus $x = 1/4$ is the only root.

4. **(D)** $3^{3(6-r)} = 3^{2(r-1)}$

    If the bases are equal, the exponents are equal.
    $$3(6 - r) = 2(r - 1)$$
    $$18 - 3r = 2r - 2$$
    $$r = 4$$

5. **(B)** The first digit can be filled in 3 ways (not zero). The second digit can be filled in 3 ways; the third digit in 2 ways and the fourth digit in 1 remaining way. Hence the number of possible integers is

$$3 \cdot 3 \cdot 2 \cdot 1 = 18.$$

6. **(E)**
$$\lim_{x \to \infty} \frac{3x^2 - \sqrt{5}x + 4}{2x^2 + 3x + \sqrt{11}}$$

$$= \lim_{x \to \infty} \frac{\frac{3x^2}{x^2} - \frac{\sqrt{5}x}{x^2} + \frac{4}{x^2}}{\frac{2x^2}{x^2} + \frac{3x}{x^2} + \frac{\sqrt{11}}{x^2}}$$

$$= \lim_{x \to \infty} \frac{3 - \frac{\sqrt{5}}{x} + \frac{4}{x^2}}{2 + \frac{3}{x} + \frac{\sqrt{11}}{x^2}}$$

$$= \frac{3}{2} = 1.50$$

7. **(E)** $x^2 - 6x + 9 + y^2 + 8y + 16 = 9 + 16$
$(x-3)^2 + (y+4)^2 = 25$
radius = $\sqrt{25}$ = 5

8. **(C)**

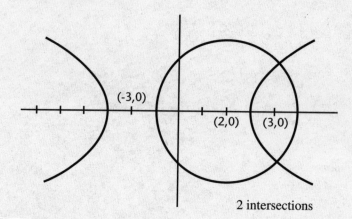

2 intersections

9. **(D)** Let each root = r
then $\quad 3r = 6$
and $\quad r = 2$

10. **(B)** One real root is 2. By DeMoivre's Theorem, the complex roots are separated by $\frac{360°}{5} = 72°$. Hence, in the second quadrant the root is $2(\cos 144° + i \sin 144°)$.

11. **(C)** $x^2 - 3x - 10 < 0$
$(x-5)(x+2) < 0$
then $x - 5 > 0$ and $x + 2 > 0$
$\quad\quad x > 5$ and $x < -2$
$\quad\quad$ Impossible
or $\quad x - 5 < 0$ and $x + 2 < 0$
$\quad\quad x < 5$ and $x > -2$
thus $\quad -2 < x < 5.$

12. **(E)**

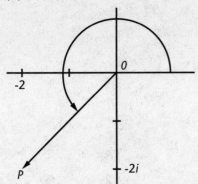

The length of OP is $2\sqrt{2}$.

The amplitude of OP is $225°$ or $\dfrac{5\pi}{4}$ radians.

Hence, the polar form is $2\sqrt{2}\left(\cos\dfrac{5\pi}{4} + i\sin\dfrac{5\pi}{4}\right)$.

13. **(C)** Since $\log_{10} 1 = 0$ and the $\log N$ decreases, it follows that $\log_{10} x < 0$.

14. **(D)** The inverse of an implication is obtained by negating the hypothesis and the conclusion. Thus the inverse of the given proposition is $p \to q$.

15. **(A)** $f(x,y) = (\ln x^2)(e^{2y})$
    $f(2, \sqrt{2}) = (\ln 4)(e^{2\sqrt{2}})$
    $= (1.386)(16.92)$
    $= 23.45$

16. **(D)** For f to be a function there must be a unique value of y for any given value of x. This is apparently true for all the sets above except where $y \geq x + 1$; in this case, for any given value of x, there is an infinity of values of y. Hence the non-function is (D).

17. **(D)** Eliminate fractions in both equations.

$$1 + xy = 2x$$
$$1 + xy = 3y$$

Hence $2x = 3y$

and $\dfrac{x}{y} = \dfrac{3}{2}$.

18. **(A)** $\lim\limits_{h \to 0} \dfrac{f(x+h) - f(x)}{h}$

$= \dfrac{3(x+h)^2 - 2(x+h) + 5 - (3x^2 - 2x + 5)}{h}$

$= \lim\limits_{h \to 0} \dfrac{3x^2 + 6xh + 3h^2 - 2x - 2h + 5 - 3x^2 + 2x - 5}{h}$

$= \lim\limits_{h \to 0} \dfrac{6xh + 3h^2 - 2h}{h} = \lim\limits_{h \to 0} 6x + 3h - 2$

$= 6x - 2$.

19. **(A)** $\log_{11} 21 = \dfrac{\log_b 21}{\log_b 11}$

$= \dfrac{\log 21}{\log 11}$ or $\dfrac{\ln 21}{\ln 11}$

$= 1.27$

20. **(B)**

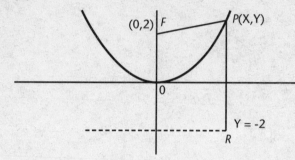

$PR = PF$
$y + 2 = \sqrt{x^2 + (y-2)^2}$
$y^2 + 4y + 4 = x^2 + y^2 - 4y + 4$
$x^2 = 8y$

21. **(C)**

Let $A = \tan^{-1} 3$,
then $\tan A = 3$
then $\sin A = \dfrac{3}{\sqrt{10}}$

22. **(D)** The graph of $y = \tan x$ indicates that the function increases in all four quadrants.

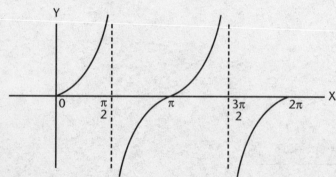

23. **(B)** Let the area of the section be A

then $\dfrac{A}{18} = \left(\dfrac{1}{3}\right)^2 = \dfrac{1}{9}$

or $A = 2$

24. **(C)** $\sin x = \dfrac{\sqrt{17}}{\sqrt{23}}$

   Since $\sin x$ and $\cos x$ are positive, x is a first quadrant angle.

   Method 1: $x = \sin^{-1}\left(\dfrac{\sqrt{17}}{\sqrt{23}}\right) = 59.29°$

   $\tan 59.29° = 1.68$

   Method 2:

   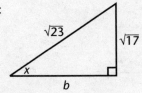

   $\left(\sqrt{17}\right)^2 + b^2 = \left(\sqrt{23}\right)^2$
   $b = \sqrt{6}$

   $\tan x = \dfrac{\text{opposite}}{\text{adjacent}} = \dfrac{\sqrt{17}}{\sqrt{6}} = 1.68$

25. **(B)** From the following diagram, we see that we can get to P by $(-5, 150°)$.

   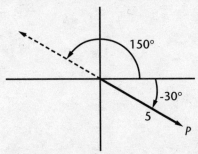

26. **(D)** $\sim p$ means two heads do not show; therefore $\sim p \land q$ means three heads show.

   The probability of 3 heads is $\dfrac{1}{2} \cdot \dfrac{1}{2} \cdot \dfrac{1}{2} = \dfrac{1}{8}$.

27. **(D)** $l = r\theta$

   $= 6 \cdot \dfrac{2\pi}{3} = 4\pi.$

28. **(D)** $\cos x (\cos x - 2) = 0$
   $\cos x = 0, \cos x = 2 \text{ (impossible)}$
   $x = 90°.$

29. **(C)** $\dfrac{\cos\theta}{\sin\theta} - \dfrac{2\cos^2\theta - 1}{\sin\theta\cos\theta}$

   $= \dfrac{\cos^2\theta - (2\cos^2\theta - 1)}{\sin\theta\cos\theta} = \dfrac{1 - \cos^2\theta}{\sin\theta\cos\theta}$

   $= \dfrac{\sin^2\theta}{\sin\theta\cos\theta} = \dfrac{\sin\theta}{\cos\theta} = \tan\theta.$

30. **(C)** The maximum value of the function is determined by the coefficient of the function. Hence maximum value is $2m$.

31. **(A)** Method 1:
$$y(t) = 2 + 4\sin t$$
$$5 = 2 + 4\sin t$$
$$\sin t = \frac{3}{4}$$
$$t = 48.6°$$
$$x(t) = 3\cos(48.6°)$$
$$= 1.98$$

or

Method 2:
$$x(t) = 3\cos t \quad \Rightarrow \quad \frac{x^2}{9} + \frac{(y-2)^2}{16} = 1$$
$$y(t) = 2 + 4\sin t$$

$$\therefore x = \pm\sqrt{9\left(1 - \frac{(y-2)^2}{16}\right)}$$

$$x = \pm\sqrt{9\left(1 - \frac{9}{16}\right)}$$

$$= 1.98$$

32. **(D)** For $x \geq 0$, $y = x - |x| = x - x = 0$
For $x \leq 0$, $y = x - |x| = x - (-x)$
or $\qquad\qquad\qquad\qquad y = 2x$

33. **(A)** Let $y = f(x) = \dfrac{1}{\sqrt{x+1}}$

then $\qquad y^2 = \dfrac{1}{x+1}$

$x + 1 = \dfrac{1}{y^2}$ or $x = \dfrac{1}{y^2} - 1 = \dfrac{1 - y^2}{y^2}$

thus $\qquad f^{-1}(x) = \dfrac{1 - x^2}{x^2}$

However, the range of $f(x)$ consists of all real numbers > 0 so that $f^{-1}(x)$ is defined for $x > 0$.

34. **(B)** By similar triangles,

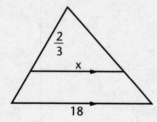

$$\frac{x^2}{18^2} = \frac{2}{3}$$
$$3x^2 = 2 \cdot 18 \cdot 18$$
$$x^2 = 2 \cdot 6 \cdot 6 \cdot 3$$
$$x = 6\sqrt{6}$$

35. **(B)** The *x*-intercepts, also called roots or zeroes, are found by setting $f(x) = 0$

$$0 = \frac{x + \sqrt{7}}{x - 1.6}$$
$$\Downarrow$$
$$x + \sqrt{7} = 0$$
$$x = -\sqrt{7}$$
$$= -2.646$$

36. **(B)** $e^x - \frac{1}{e^x} + 1 = 0$

Multiply by e^x: $e^{2x} + e^x - 1 = 0$

By quadratic formula: $e^x = \frac{-1 \pm \sqrt{5}}{2}$

Since e^x can only be a positive number, reject the minus sign in the formula.

Thus $e^x = \frac{\sqrt{5} - 1}{2}$

only 1 root

37. **(C)** When $x \geq 0$ and $y \geq 0$,
    the equation is $x + y = 4$.
    When $x \leq 0$ and $y \leq 0$,
    the equation is $-x - y = 4$.
    When $x \leq 0$ and $y \geq 0$,
    the equation is $-x + y = 4$.
    When $x \geq 0$ and $y \leq 0$,
    the equation is $x - y = 4$.
    The graph thus becomes the following:

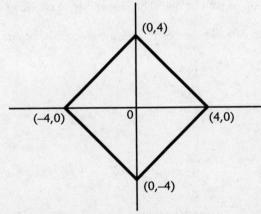

The graph consists of the sides of a square.

38. **(C)** From the right triangle,

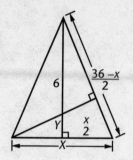

$$\left(\frac{36-x}{2}\right)^2 = \frac{x^2}{4} + 36$$
$$(36-x)^2 = x^2 + 144$$
$$x = 16$$
$$\frac{36-x}{2} = 10$$

By equating areas, we get
$$10y = 6 \cdot 16 = 96$$
$$y = 9.6$$

39. **(D)** Let the original radius = 1.

Then the original volume $= \frac{4}{3}\pi \cdot 1^1 = \frac{4}{3}\pi$

The new radius = 2.

The new volume $= \frac{4}{3}\pi \cdot 2^3 = \frac{4}{3}\pi \cdot 8$

The increase in volume is $\frac{4}{3}\pi(8-1) = \frac{4}{3}\pi \cdot 7$

The percent increase $= \dfrac{\frac{4}{3}\pi \cdot 7}{\frac{4}{3}\pi} \times 100$

$= 700.$

40. **(E)** The two numbers must both be even or both be odd. The probability of choosing two even numbers is $\frac{6}{12} \cdot \frac{5}{11} = \frac{5}{22}$: the same for two odd numbers, hence the probability of one or the other is $\frac{5}{22} + \frac{5}{22} = \frac{5}{11}$.

41. **(D)** $\quad \dfrac{3}{\sec^2 x} = 5 - \dfrac{3}{\csc^2 x}$

$$3\cos^2 x = 5 - 3\sin^2 x$$
$$3(\cos^2 x + \sin^2 x) = 5$$
$$3 \neq 5$$

Hence no value of x.

42. **(C)** Let the series be $a, ar, ar^2, \ldots, ar^{12}$.

Then $a = m^{1/3}$ and $ar^2 = m^{1/2}$

$$m^{1/3}r^2 = m^{1/2}$$

$$r^2 = m^{1/2 - 1/3} = m^{1/6}$$

$$r = m^{1/12}$$

Then $\quad ar^{12} = m^{1/3}(m^{1/12})^{12} = m^{4/3}$.

43. **(E)**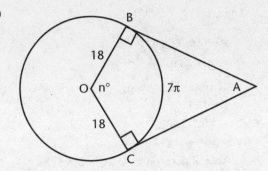

Let $\angle BOC = n°$

then $7\pi = \dfrac{n}{360} \cdot 2\pi \cdot 18$

$7\pi = \dfrac{n\pi}{10}$

$n = 70°$

$\angle BAC = 180° - 70° = 110°$.

44. **(C)**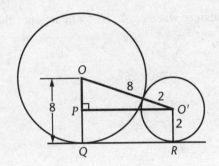

$OO' = 8 + 2 = 10$
$OP = 8 - 2 = 6$
Hence $O'P = 8$
Thus $QR = 8$

45. **(C)** At the y-intercept $x = 0$

$$-\dfrac{0^2}{9} + \dfrac{y^2}{11} = 1$$

$$y^2 = 11$$

$$y = \pm\sqrt{11} = \pm 3.32$$

46. **(E)** The remainder is $5(-1)^{2n+1} - 10(-1)^{2n} + 3(-1)^{2n-1} + 5$.
Since $2n + 1$ and $2n - 1$ are odd and $2n$ is even, the remainder
$= 5(-1) - 10(1) + 3(-1) + 5$
$= -5 - 10 - 3 + 5 = -13$.

47. **(B)** $r_1 + r_2 = p$ and $r_1 r_2 = q$.
Then $(r_1 + r_2)^2 = p^2$ or $r_1^2 + r_2^2 + 2r_1 r_2 = p^2$.
Thus $r_1^2 + r_2^2 = p^2 - 2q$.

48. **(E)** $\dfrac{2^{n+4} - 2^{n+1}}{2^{n+4}} = 1 - 2^{-3}$

$= 1 - \dfrac{1}{2^3} = 1 - \dfrac{1}{8}$

$= \dfrac{7}{8}$.

49. **(D)** 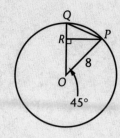 The area = $8 \times \triangle OPQ$

$\triangle OPQ = \frac{1}{2} OQ \times PR$

$OQ = OP = 8$

Since OPR is a right isosceles $\triangle$,

$PR = \frac{8}{\sqrt{2}} = 4\sqrt{2}$.

Area of octagon = $4 \cdot 8 \cdot 4\sqrt{2} = 128\sqrt{2}$.

50. **(B)** The maximum value for $3 - \cos 2x$ is attained when $\cos 2x$ is at a minimum value. Since the least value for the cosine of an angle is -1,

let $\cos 2x = -1$
then $2x = \pi$
$x = \frac{\pi}{2}$.

Sample Test 3
Answer Sheet

Math Level IIC

1 Ⓐ Ⓑ Ⓒ Ⓓ Ⓔ	11 Ⓐ Ⓑ Ⓒ Ⓓ Ⓔ	21 Ⓐ Ⓑ Ⓒ Ⓓ Ⓔ	31 Ⓐ Ⓑ Ⓒ Ⓓ Ⓔ	41 Ⓐ Ⓑ Ⓒ Ⓓ Ⓔ
2 Ⓐ Ⓑ Ⓒ Ⓓ Ⓔ	12 Ⓐ Ⓑ Ⓒ Ⓓ Ⓔ	22 Ⓐ Ⓑ Ⓒ Ⓓ Ⓔ	32 Ⓐ Ⓑ Ⓒ Ⓓ Ⓔ	42 Ⓐ Ⓑ Ⓒ Ⓓ Ⓔ
3 Ⓐ Ⓑ Ⓒ Ⓓ Ⓔ	13 Ⓐ Ⓑ Ⓒ Ⓓ Ⓔ	23 Ⓐ Ⓑ Ⓒ Ⓓ Ⓔ	33 Ⓐ Ⓑ Ⓒ Ⓓ Ⓔ	43 Ⓐ Ⓑ Ⓒ Ⓓ Ⓔ
4 Ⓐ Ⓑ Ⓒ Ⓓ Ⓔ	14 Ⓐ Ⓑ Ⓒ Ⓓ Ⓔ	24 Ⓐ Ⓑ Ⓒ Ⓓ Ⓔ	34 Ⓐ Ⓑ Ⓒ Ⓓ Ⓔ	44 Ⓐ Ⓑ Ⓒ Ⓓ Ⓔ
5 Ⓐ Ⓑ Ⓒ Ⓓ Ⓔ	15 Ⓐ Ⓑ Ⓒ Ⓓ Ⓔ	25 Ⓐ Ⓑ Ⓒ Ⓓ Ⓔ	35 Ⓐ Ⓑ Ⓒ Ⓓ Ⓔ	45 Ⓐ Ⓑ Ⓒ Ⓓ Ⓔ
6 Ⓐ Ⓑ Ⓒ Ⓓ Ⓔ	16 Ⓐ Ⓑ Ⓒ Ⓓ Ⓔ	26 Ⓐ Ⓑ Ⓒ Ⓓ Ⓔ	36 Ⓐ Ⓑ Ⓒ Ⓓ Ⓔ	46 Ⓐ Ⓑ Ⓒ Ⓓ Ⓔ
7 Ⓐ Ⓑ Ⓒ Ⓓ Ⓔ	17 Ⓐ Ⓑ Ⓒ Ⓓ Ⓔ	27 Ⓐ Ⓑ Ⓒ Ⓓ Ⓔ	37 Ⓐ Ⓑ Ⓒ Ⓓ Ⓔ	47 Ⓐ Ⓑ Ⓒ Ⓓ Ⓔ
8 Ⓐ Ⓑ Ⓒ Ⓓ Ⓔ	18 Ⓐ Ⓑ Ⓒ Ⓓ Ⓔ	28 Ⓐ Ⓑ Ⓒ Ⓓ Ⓔ	38 Ⓐ Ⓑ Ⓒ Ⓓ Ⓔ	48 Ⓐ Ⓑ Ⓒ Ⓓ Ⓔ
9 Ⓐ Ⓑ Ⓒ Ⓓ Ⓔ	19 Ⓐ Ⓑ Ⓒ Ⓓ Ⓔ	29 Ⓐ Ⓑ Ⓒ Ⓓ Ⓔ	39 Ⓐ Ⓑ Ⓒ Ⓓ Ⓔ	49 Ⓐ Ⓑ Ⓒ Ⓓ Ⓔ
10 Ⓐ Ⓑ Ⓒ Ⓓ Ⓔ	20 Ⓐ Ⓑ Ⓒ Ⓓ Ⓔ	30 Ⓐ Ⓑ Ⓒ Ⓓ Ⓔ	40 Ⓐ Ⓑ Ⓒ Ⓓ Ⓔ	50 Ⓐ Ⓑ Ⓒ Ⓓ Ⓔ

Directions: For each question in the sample test, select the best of the answer choices and blacken the corresponding space on this answer sheet.

Please note: (a) You will need to use a calculator in order to answer some, though not all, of the questions in this test. As you look at each question, you must decide whether or not you need a calculator for the specific question. A four-function calculator is not sufficient; your calculator must be at least a scientific calculator. Calculators that can display graphs and programmable calculators are also permitted.

(b) Set your calculator to radian mode or degree mode depending on the requirements of the question.

(c) All figures are accurately drawn and are intended to supply useful information for solving the problems that they accompany. Figures are drawn to scale UNLESS it is specifically stated that a figure is not drawn to scale. Unless otherwise indicated, all figures lie in a plane.

(d) The domain of any function f is assumed to be the set of all real numbers x for which $f(x)$ is a real number except when this is specified not to be the case.

(e) Use the reference data below as needed.

REFERENCE DATA

Solid	Volume	Other		
Right circular cone	$V = \frac{1}{3}\pi r^2 h$	$S = cl$	V = volume r = radius h = height	S = lateral area c = circumference of base l = slant height
Sphere	$V = \frac{4}{3}\pi r^3$	$S = 4\pi r^2$	V = volume r = radius S = surface area	
Pyramid	$V = \frac{1}{3}Bh$		V = volume B = area of base h = height	

Sample Test 3

MATH LEVEL IIC

50 Questions • Time—60 Minutes

DO YOUR FIGURING HERE.

1. If x and y are real numbers, which one of the following relations is a function of x?

   (A) $\{(x, y) \mid x = y^2 - 1\}$ (B) $\{(x, y) \mid y = x^2 + 1\}$
   (C) $\{(x, y) \mid y = \pm\sqrt{16 - x^2}\}$ (D) $\{(x, y) \mid y < x - 2\}$
   (E) $x = \sin y$

2. If P represents the set of rhombi and Q the set of rectangles, then the set $P \cap Q$ represents the set of

   (A) squares   (B) trapezoids   (C) parallelograms
   (D) quadrilaterals   (E) rectangles

3. The length of the vector that could correctly be used to represent in the complex plane the number $3 - i\sqrt{2}$ is:

   (A) 11   (B) $\sqrt{11}$ (C) $3\sqrt{2}$ (D) $\sqrt{5}$ (E) $\sqrt{13}$

4. If $\log_8 p = 2.5$ and $\log_2 q = 5$, then p expressed in terms of q is:

   (A) $p = q^2$ (B) $p = q^8$ (C) $p = q^{3/2}$ (D) $p = q^{5/2}$ (E) $p = q^3$

5. If circle R, of area 4 square inches, passes through the center of, and is tangent to circle S, then the area of circle S, in square inches, is:

   (A) 8   (B) $8\sqrt{2}$ (C) $16\sqrt{2}$ (D) 12   (E) 16

6. In a cube, the ratio of the longest diagonal to a diagonal of a base is:

   (A) $\sqrt{2}:\sqrt{3}$ (B) $\sqrt{6}:2$ (C) $\sqrt{6}:\sqrt{2}$ (D) $\sqrt{3}:\sqrt{2}$ (E) 2:1

7. Which ordered number pair represents the center of the circle, $x^2 + y^2 - 6x + 4y - 12 = 0$?

   (A) (9,4)   (B) (–3,2)   (C) (3,–2)   (D) (–6,4)   (E) (6,4)

8. $\sin(135° + x) + \sin(135° - x)$ equals

   (A) $\sqrt{2}\sin x$ (B) –1   (C) $\sqrt{3}\cos x$ (D) $\sqrt{2}\cos x$
   (E) $\sqrt{3}\sin x$

9. If $3^7 = 7^x$, what is the value of x?

   (A) .95
   (B) 1.95
   (C) 2.95
   (D) 3.95
   (E) 4.15

10. In a circle a central angle of 60° intercepts an arc of 15 inches. How many inches in the radius of the circle?

    (A) $\dfrac{45}{\pi}$
    (B) $\dfrac{\pi}{5}$
    (C) 4
    (D) $\dfrac{2\pi}{3}$
    (E) not computable from given data

11. For what value of x is the function $\sin x \cos x$ a maximum?

    (A) 2π (B) $\dfrac{3\pi}{4}$ (C) $\dfrac{\pi}{2}$ (D) $\dfrac{2\pi}{3}$ (E) $\dfrac{\pi}{4}$

12. Four men line up in a row. What is the probability that a certain two are next to each other?

    (A) $\dfrac{1}{6}$ (B) $\dfrac{1}{4}$ (C) $\dfrac{1}{3}$ (D) $\dfrac{1}{2}$ (E) $\dfrac{2}{3}$

13. What is $\lim\limits_{x \to \infty} \dfrac{\sqrt{7}x^2 + 3x - 2}{x^2 + 5}$?

    (A) 2.32
    (B) 2.43
    (C) 2.54
    (D) 2.65
    (E) 2.76

14. If P implies Q, an equivalent statement is

    (A) Q implies P
    (B) Q is a necessary condition for P
    (C) P is a necessary condition for Q
    (D) Not P implies Q
    (E) Not P implies not Q

DO YOUR FIGURING HERE.

15. T varies directly as the square of r and inversely as the cube of s. If r is tripled and s is doubled, the value of T is

   (A) multiplied by 3/2

   (B) multiplied by 6

   (C) multiplied by 9/8

   (D) multiplied by 2

   (E) divided by 2

16. If $\log x \geq \log 2 + \frac{1}{2} \log x$, then

   (A) $x \geq 2$ (B) $x \leq 2$ (C) $x \leq 4$ (D) $x \geq 4$
   (E) $x \geq 1$

17. $2 \cos^3 A \sin A + 2 \sin^3 A \cos A$ equals which one of the following?

   (A) $\cos 2A$ (B) $2 \sin A$ (C) $2 \cos A$ (D) $\cos^2 A$
   (E) $\sin 2A$

18. An isosceles triangle with base 24 and legs of 15 is inscribed in a circle. Find the radius.

   (A) 7   (B) $12\frac{1}{2}$ (C) 25   (D) $25\frac{1}{2}$
   (E) cannot be determined

19. Two roots of $4x^3 + 8x^2 + Kx - 18 = 0$ are equal numerically but opposite in sign. Find the value of K.

   (A) -2 (B) $+2$ (C) -9 (D) $+9$ (E) $-\frac{9}{2}$

20. If $\log_{10} x = y$ and $\log_e 10 = \frac{1}{m}$, then

   (A) $\log_e x = \frac{y}{m}$ (B) $\log_e x = \frac{m}{y}$ (C) $\log_e y = \frac{m}{x}$

   (D) $\log_e y = \frac{x}{m}$ (E) none of these

21. Find the angle between π and 2π that satisfies the equation:
    $2 \sin^2 x + 5 \cos x + 1 = 0$

   (A) $\frac{7\pi}{6}$ (B) $\frac{3\pi}{2}$ (C) $\frac{5\pi}{3}$ (D) $\frac{11\pi}{6}$ (E) $\frac{4\pi}{3}$

22. Two heavenly bodies are respectively 8.9×10^8 miles and 2.7×10^7 miles distant from the earth. How much farther away from the earth is the first than the second?

   (A) 6.2×10^7 (B) 8.6×10^7
   (C) 8.6×10^8 (D) 863×10^6
   (E) 62×10^8

23. Find the product of an infinite number of terms:

    $3^{1/2} \times 3^{1/4} \times 3^{1/8} \times 3^{1/16} \times \ldots$

    (A) $3^{1/2}$ (B) $\sqrt[4]{3}$ (C) 3^2 (D) 3   (E) 1

24. Which of the following is an (x,y) coordinate pair located on the ellipse $4x^2 + 9y^2 = 100$?

    (A) (1, 3.5)
    (B) (1.4, 3.2)
    (C) (1.9, 2.9)
    (D) (2.3, 3.1)
    (E) (2.7, 2.6)

25. If $\log_4 44 = x$, what is the value of x?

    (A) 2.73
    (B) 2.70
    (C) 1.67
    (D) 1.64
    (E) 1.61

26. A boy walks diagonally across a square lot. What percent does he save by not walking along the edges (approximately)?

    (A) 22   (B) 29   (C) 33   (D) 20   (E) 24

27. What is the largest rod that can just fit into a box 24" × 8" × 6" (in inches)?

    (A) 24   (B) 25   (C) 26   (D) 28   (E) 30

28. Find the solution set of the inequality $x^2 - x - 6 < 0$.

    (A) $x > -2$ (B) $-2 < x < 3$
    (C) $x > 3$ and $x < -2$ (D) $x > 3$ or $x < -2$
    　　　　　　(E) $x < 3$

DO YOUR FIGURING HERE.

29. If $f(x) = x^2 + 2$ and $g(x) = 1 - \frac{1}{x}$, write the expression $g[f(x)]$ in terms of x.

    (A) $1 - \frac{1}{x^2}$
    (B) $\frac{x}{x^2 - 1}$
    (C) $\frac{x^2 + 2}{x^2 + 1}$
    (D) $\frac{x^2 + 1}{x^2 + 2}$
    (E) none of these

DO YOUR FIGURING HERE.

30. Find the value of K if $(x + 1)$ is a factor of $x^8 + Kx^3 - 2x + 1$.

    (A) 1   (B) 2   (C) 3   (D) 4   (E) 5

31. The portion of the plane, whose equation is
    $$8x + 20y + 15z = 120,$$
    that lies in the first octant forms a pyramid with the coordinate planes. Find its volume.

    (A) 90
    (B) 120
    (C) 140
    (D) 150
    (E) 160

32. Write $\left[\sqrt{2}(\cos 30° + i \sin 30°)\right]^2$ in the form $a + bi$.

    (A) $1 + \sqrt{3}i$
    (B) $1 - \sqrt{3}i$
    (C) $\frac{3}{2} + \frac{1}{2}i$
    (D) $\frac{3}{2} - \frac{1}{2}i$
    (E) $\sqrt{3}i$

33. Let the symbol Δ be defined as
    $a \Delta b = \sin a \sin b - \cos a \cos b$. What is the value of $17° \Delta 33°$?

    (A) .766
    (B) .643
    (C) .125
    (D) 0
    (E) −.643

34. What is the equation of the perpendicular bisector of the line segment whose end points are $(2, 6)$ and $(-4, 3)$?

    (A) $4x + 2y - 5 = 0$
    (B) $4x - 2y + 13 = 0$
    (C) $x - 2y + 10 = 0$
    (D) $x + 2y - 8 = 0$
    (E) $4x + 2y - 13 = 0$

GO ON TO THE NEXT PAGE

35. Of the following, the one which is a cube root of *i* is:

    (A) +*i*
    (B) cos 30° + *i* sin 30°
    (C) cos 90° + *i* sin 90°
    (D) cos 120° + *i* sin 120°
    (E) cos 60° + *i* sin 60°

36. What is the period of the curve whose equation is

    $$y = \frac{1}{3}\left(\cos^2 x - \sin^2 x\right)?$$

    (A) 60° (B) 120°
    (C) 180° (D) 720°
    (E) $\frac{1}{3}$

37. In figure 37, what is the area of triangle *NJL*?

    (A) 11.79
    (B) 13.85
    (C) 15.31
    (D) 17.10
    (E) 17.82

    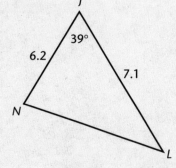

    Fig. 37

38. A right circular cylinder is circumscribed about a sphere. If *S* represents the surface area of the sphere and *T* represents the total area of the cylinder, then

    (A) $S = \frac{2}{3}T$ (B) $S < \frac{2}{3}T$
    (C) $S > \frac{2}{3}T$ (D) $S \leq \frac{2}{3}T$
    (E) $S \geq \frac{2}{3}T$

39. If $|x-2|<5$, what are the possible values of x?

   (A) $0<x<5$
   (B) $0<x<2$
   (C) $-3<x\leq 7$
   (D) $-3\leq x<7$
   (E) $-3<x<7$

40. How many even numbers greater than 40,000 may be formed using the digits 3, 4, 5, 6, and 9, if each digit must be used exactly once in each number?

   (A) 36   (B) 48   (C) 64   (D) 96   (E) 112

41. A regular octagon is formed by cutting off each corner of a square whose side is 6. Find the length of one side of the octagon.

   (A) 2
   (B) $2\sqrt{2}$
   (C) $2\sqrt{2}-2$
   (D) $6\sqrt{2}-6$
   (E) $\sqrt{2}-1$

42. What is $\lim_{x\to\sqrt{7}}\dfrac{x^3-3x+1}{x+1}$?

   (A) 1.42
   (B) 3.17
   (C) 5.38
   (D) 7.00
   (E) 8.67

43. All triangles in the set of triangles having a given side and a given angle opposite that side

   (A) are congruent
   (B) are similar
   (C) are equivalent
   (D) have the same inscribed circle
   (E) have the same circumscribed circle

44. Find the value of $\log_8 \sqrt[3]{0.25}$

   (A) $\dfrac{2}{3}$
   (B) $-\dfrac{2}{9}$
   (C) $-\dfrac{4}{3}$
   (D) $\dfrac{3}{4}$
   (E) $-\dfrac{3}{2}$

GO ON TO THE NEXT PAGE

45. For what positive value of m will the line $y = mx + 5$ be tangent to the circle $x^2 + y^2 = 9$?

   (A) 1 (B) 2 (C) $\frac{2}{3}$ (D) $\frac{3}{4}$ (E) $\frac{4}{3}$

46. The graph of the curve whose parametric equations are $x = a \sin t$ and $y = b \cos t$ is a(n):

   (A) ellipse
   (B) circle
   (C) parabola
   (D) hyperbola
   (E) straight line

47. The graph of $y = |x - 2| + 2$ consists of

   (A) one straight line
   (B) a pair of straight line rays
   (C) the sides of a square
   (D) a circle
   (E) a parabola

48. The converse of $\sim p \to q$ is equivalent to

   (A) $p \to \sim q$
   (B) $p \to q$
   (C) $\sim q \to p$
   (D) $q \to p$
   (E) $\sim p \to \sim q$

49. Write an equation of lowest degree, with real coefficients, if two of its roots are -1 and $1 + i$.

   (A) $x^3 + x^2 + 2 = 0$
   (B) $x^3 - x^2 - 2 = 0$
   (C) $x^3 - x + 2 = 0$
   (D) $x^3 - x^2 + 2 = 0$
   (E) none of these

50. If $\log_r 6 = S$ and $\log_r 3 = T$, then $\log_r \left(\frac{r}{2}\right)$ is equal to

   (A) $\frac{1}{2} \log_2 r$ for any r
   (B) $1 - S + T$
   (C) $1 - S - T$
   (D) $\log_r 2 - 1$
   (E) zero, if $r = 4$

STOP

Sample Test 3
Answer Key

Math Level IIC

1. B	11. E	21. E	31. B	41. D
2. A	12. D	22. C	32. A	42. C
3. B	13. D	23. D	33. E	43. E
4. C	14. B	24. B	34. A	44. B
5. E	15. C	25. A	35. B	45. E
6. D	16. D	26. B	36. C	46. A
7. C	17. E	27. C	37. B	47. B
8. D	18. B	28. B	38. A	48. A
9. D	19. C	29. D	39. E	49. D
10. A	20. A	30. D	40. A	50. B

Solutions

1. **(B)** For y to be a function of x, there must be a unique value of y for any given value of x. This would be true only for $y = x^2 + 1$. In A, C, D, and E, y may take on 2 or more values for a given x.

2. **(A)** The set $P \cap Q$ includes as elements all rectangles that are also rhombi. These elements make up the set of squares.

3. **(B)** length $= \sqrt{3^2 + \sqrt{2}^2}$
   $\sqrt{9+2} = \sqrt{11}$

4. **(C)** $8^{2.5} = p$ and $2^5 = q$
   then $(2^3)^{2.5} = p$ or $p = 2^{7.5} = (2^5)^{1.5}$
   thus $p = (2^5)^{3/2} = q^{3/2}$.

5. **(E)** R is internally tangent to S and its diameter is half that of S. Hence S has an area 4 times that of r or 16 square inches.

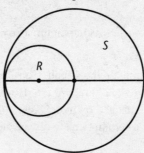

6. **(D)** Consider a cube of edge 1.
   Then the longest diagonal $= \sqrt{1^2 + 1^2 + 1^2} = \sqrt{3}$
   The diagonal of a base $= \sqrt{1^2 + 1^2} = \sqrt{2}$
   Hence the ratio $= \sqrt{3} : \sqrt{2}$.

7. **(C)** $x^2 - 6x + 9 + y^2 + 4y + 4 = 12 + 13$
   $(x - 3)^2 + (y + 2)^2 = 25$
   Center is $(3, -2)$

8. **(D)** $\sin 135° \cos x + \cos 135° \sin x$
   $\qquad + \sin 135° \cos x - \cos 135° \sin x$
   $= 2 \sin 135° \cos x$
   $= 2 \dfrac{\sqrt{2}}{2} \cos x$
   $= \sqrt{2} \cos x.$

9. **(D)**
   $7^x = 3^7$
   $x \ln 7 = 7 \ln 3$
   $x = \dfrac{7 \ln 3}{\ln 7}$
   $x = \dfrac{7(1.099)}{(1.946)}$
   $x = 3.95$

10. **(A)** $l = r\theta$
    $15 = \dfrac{\pi}{3} r$
    $r = \dfrac{45}{\pi}$

11. **(E)** Consider $y = \sin x \cos x$
    $= \dfrac{1}{2} \sin 2x$
    $\sin 2x$ has a period of $180°$ and reaches its maximum at $x = 45°$ or $\dfrac{\pi}{4}$.

12. **(D)** There are $4! = 24$ ways of lining up the 4 men. Consider the certain two as one unit and determine the number of ways of lining up three; $3! = 6$. However the two men can be next to each other in twice as many ways as indicated by merely switching places. Hence, there are 12 ways of the four men lining up so that a certain two are always next to each other.
    Probability $= \dfrac{12}{24} = \dfrac{1}{2}$.

13. **(D)** $$\lim_{x\to\infty} \frac{\sqrt{7}x^2+3x-2}{x^2+5}$$

$$= \lim_{x\to\infty} \frac{\frac{\sqrt{7}x^2}{x^2}+\frac{3x}{x^2}-\frac{2}{x^2}}{\frac{x^2}{x^2}+\frac{5}{x^2}}$$

$$= \lim_{x\to\infty} \frac{\sqrt{7}+\cancelto{0}{\frac{3}{x}}-\cancelto{0}{\frac{2}{x^2}}}{1+\cancelto{0}{\frac{5}{x^2}}} = \sqrt{7}$$

$$= 2.64575$$

14. **(B)** The converse, inverse, and negative are not equivalent statements. The contrapositive, $\sim Q \to \sim P$, is equivalent and this is the same as saying that Q is a necessary condition for P.

15. **(C)** $T = \dfrac{Kr^2}{s^3}$

$$T' = \frac{K(3r)^2}{(2s)^3} = \frac{9Kr^2}{8s^3} = \frac{9}{8} \cdot \frac{Kr^2}{s^3}$$

Hence, $T' = \dfrac{9}{8}T$.

16. **(D)** $\log x \geq \log 2 + \log x^{1/2}$

$\log x - \log x^{1/2} \geq \log 2$

$\log \dfrac{x}{x^{1/2}} \geq \log 2$

$x^{1/2} \geq 2$

$x \geq 4$

17. **(E)** $2\cos^3 A \sin A + 2\sin^3 A \cos A$
    $= 2\sin A \cos A (\cos^2 A + \sin^2 A)$
    $= 2\sin A \cos A = \sin 2A.$

18. **(B)**

$r^2 = 12^2 + (r-9)^2$
$r^2 = 144 + r^2 - 18r + 81$
$18r = 225$
$r = 12\dfrac{1}{2}$

19. **(C)** Let the roots be $r, -r,$ and s

    then $r - r + s = -2$ and $s = -2$.

    Also $-2(r)(-r) = \frac{9}{2}$ or $r^2 = \frac{9}{4}$ and $r = \pm \frac{3}{2}$

    $rs - rs - r^2 = -r^2 = \frac{K}{4}$ or $K = -4r^2$

    Thus $K = -9$.

20. **(A)** Since $\log_{10} x = y$, $10^y = x$

    From $\log_e 10 = \frac{1}{m}$, $e^{1/m} = 10$

    Substituting, we get

    $(e^{1/m})^y = x$

    $e^{y/m} = x$

    or $\log_e x = \frac{y}{m}$

21. **(E)** $2(1 - \cos^2 x) + 5 \cos x + 1 = 0$

    $2 - 2\cos^2 x + 5 \cos x + 1 = 0$

    $2 \cos^2 x - 5 \cos x - 3 = 0$

    $(2 \cos x + 1)(\cos x - 3) = 0$

    $\cos x = -\frac{1}{2}$, $\cos x = 3$ (impossible)

    $x = 240°$ or $\frac{4\pi}{3}$.

22. **(C)** $8.9 \times 10^8 - 2.7 + 10^7$

    $= 10^7 (8.9 + 10 - 2.7)$

    $= 10^7 (86.3)$

    Writing this to two significant figures, we get 8.6×10^8.

23. **(D)** $3^{1/2} \times 3^{1/4} \times 3^{1/8} \times 3^{1/16} \times \ldots$

    $= 3^{1/2 + 1/4 + 1/8 + 1/16 + \ldots}$

    The exponent is the sum of an infinite geometric series

    $S = \frac{a}{1-r} = \frac{1/2}{1 - 1/2} = 1$

    Thus $3^1 = 3$.

24. **(B)**
$$4x^2 + 9y^2 = 100$$
$$\Downarrow$$
$$9y^2 = 100 - 4x^2$$
$$y = \pm\sqrt{\frac{100-4x^2}{9}}$$

x	y
1	3.27
1.4	3.20
1.9	3.08
2.3	2.96
2.7	2.81

25. **(A)** $\log_4 44 = x$

$$4^x = 44$$
$$x \log 4 = \log 44$$
$$x = \frac{\log 44}{\log 4} = \frac{(1.643)}{(.6021)}$$
$$= 2.73$$

26. **(B)** He saves $(2x - x\sqrt{2})$.

Percent saved $= \frac{2x - x\sqrt{2}}{2x} \cdot 100$

$= \frac{2-\sqrt{2}}{2} \cdot 100 = (2-\sqrt{2})50$

$= 50(2 - 1.414)$

$= 50 \times .586$

$= 29.3$

$= 29$

27. **(C)** Longest dimension is along the diagonal.
$d^2 = 24^2 + 8^2 + 6^2$
$d^2 = 576 + 64 + 36 = 676$
$d = \sqrt{676} = 26$

28. **(B)** $x^2 - x - 6 < 0$
$\quad (x-3)(x+2) < 0$

Either $x - 3 < 0$ and $x + 2 > 0$ or $x - 3 > 0$ and $x + 2 > 0$
$x < 3$ and $x > -2$ or $x > 3$ and $x < -2$
$-2 < x < 3$ \qquad\qquad this is impossible

29. **(D)** $g[f(x)] = 1 - \dfrac{1}{f(x)} = 1 - \dfrac{1}{x^2 + 2}$

$= \dfrac{x^2 + 2 - 1}{x^2 + 2} = \dfrac{x^2 + 1}{x^2 + 2}$

30. **(D)** Let $P(x) = x^8 + Kx^3 - 2x + 1$
then $P(-1) = (-1)^8 + K(-1)^3 - 2(-1) + 1 = 0$
or $\qquad\qquad\qquad 1 - K + 2 + 1 = 0$
$\qquad\qquad\qquad\qquad\qquad K = 4$

31. **(B)**

The x-intercept is given by $8x = 120$
or $\qquad\qquad\qquad\qquad x = 15$
the y-intercept: $20y = 120$
$\qquad\qquad\qquad\qquad y = 6$
the z-intercept: $15z = 120$
$\qquad\qquad\qquad\qquad z = 8$
The area of the base = $\dfrac{1}{2} \cdot 15 \cdot 6 = 45$.
Volume = $\dfrac{1}{3} \cdot 45 \cdot 8 = 120$.

32. **(A)** By De Moivre's Theorem
$[\sqrt{2}(\cos 30° + i \sin 30°)]^2 = 2(\cos 60° + i \sin 60°)$.

$= 2\left(\dfrac{1}{2} + \dfrac{\sqrt{3}}{2}i\right)$

$= 1 + \sqrt{3}i$

33. **(E)** Method 1: $\qquad\qquad 17° \triangle 33°$
$= \sin 17° \sin 33° - \cos 17° \cos 33°$
$= (.2924)(.5446) - (.9563)(.8387)$
$= -.643$

Method 2 $\qquad \cos(a + b) = \cos a \cos b - \sin a \sin b$
$a \triangle b = -\cos(a + b)$
$17° \triangle 33° = -\cos(17 + 33)$
$= -\cos(50°)$
$= -.643$

34. **(A)**
$$(x-2)^2 + (y-6)^2 = (x+4)^2 + (y-3)^2$$
$$x^2 - 4x + 4 + y^2 - 12y + 36 = x^2 + 8x + 16 + y^2 - 6y + 9$$
$$-12x - 6y + 15 = 0$$
$$4x + 2y - 5 = 0$$

35. **(B)** $i = \cos 90° + i \sin 90°$
$i^{1/3} = (\cos 90° + i \sin 90°)^{1/3}$
　　$= \cos 30° + i \sin 30°$　　　1st root
　　$= \cos 150° + i \sin 150°$　　2nd root
　　$= \cos 270° + i \sin 270°$　　3rd root

36. **(C)** $y = \frac{1}{3}(\cos^2 x - \sin^2 x) = \frac{1}{3}\cos 2x$
Since the period of $\cos x$ is 360°, the period of $\cos 2x$ is 180°.

37. **(B)** $A = \frac{1}{2}ab \sin C$
$= \frac{1}{2}(\overline{NJ})(\overline{JL})\sin J$
$= \frac{1}{2}(6.2)(7.1)\sin 39°$
$= 13.851$

38. **(A)**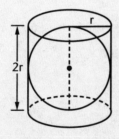
$S = 4\pi r^2$
$T = 2\pi r^2 + 2\pi r(2r)$
　$= 6\pi r^2$
$\frac{S}{T} = \frac{4\pi r^2}{6\pi r^2} = \frac{2}{3}$
$S = \frac{2}{3}T$

39. **(E)** If $x - 2 > 0$ or $x > 2$, then $x - 2 < 5$
or　$x < 7$
If　$x - 2 < 0$ or $x < 2$, then $-(x - 2) < 5$
or　$x - 2 > -5$
or　$x > -3$
Thus, $-3 < x < 7$
or　$|x - 2| < 5$
　　$-5 < x - 2 < 5$
Adding 2 to all three terms of inequality,
　　$-3 < x < 7$

40. **(A)** The last digit may only be filled by 4 or 6, thus in 2 ways. This leaves 3 remaining numbers for the first digit, 3 more for the second digit, 2 for the third digit and 1 for the fourth digit. $3 \cdot 3 \cdot 2 \cdot 1 \cdot 2 = 36$.

41. **(D)**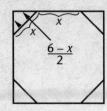

$$x^2 = \left(\frac{6-x}{2}\right)^2 + \left(\frac{6-x}{2}\right)^2$$

$$x^2 = \frac{(6-x)^2}{2}$$

$$2x^2 = 36 - 12x + x^2$$

$$x^2 + 12x - 36 = 0$$

$$x = -\frac{12 \pm \sqrt{288}}{2}$$

Reject the negative square root

$$x = -\frac{12 \pm 12\sqrt{2}}{2} = 6\sqrt{2} - 6.$$

42. **(C)** $\lim_{x \to \sqrt{7}} \frac{x^3 - 3x + 1}{x + 1}$

$$= \frac{(\sqrt{7})^3 - 3(\sqrt{7}) + 1}{(\sqrt{7}) + 1}$$

$$= 5.378$$

43. **(E)** If the given side is p and the angle opposite is P, then the diameter, d, of the circumscribed circle is given by

$$d = \frac{p}{\sin P}$$

Hence, all circumscribed circles have the same diameter. Thus all triangles have the same circumscribed circle.

44. **(B)** $\log_8 \sqrt[3]{0.25} = \frac{1}{3} \log_8 \frac{1}{4}$

Let $\log_8 \frac{1}{4} = x$

thus $8^x = \frac{1}{4}$ or $2^{3x} = 2^{-2}$

Hence $3x = -2$ and $x = -\frac{2}{3}$

$$\frac{1}{3}\left(-\frac{2}{3}\right) = -\frac{2}{9}$$

45. **(E)**
$$x^2 + y^2 = 9$$
$$x^2 + (mx + 5)^2 = 9$$
$$x^2 + m^2 x^2 + 10mx + 25 - 9 = 0$$
$$(1 + m^2)x^2 + 10mx + 16 = 0$$

If the line is tangent to the circle, the discriminant of the quadratic must be zero.

$$100m^2 - 64(1 + m^2) = 0$$
$$36m^2 = 64$$
$$m = \frac{4}{3}$$

46. **(A)** $\dfrac{x}{a} = \sin t,\ \dfrac{y}{b} = \cos t$

$\dfrac{x^2}{a^2} = \sin^2 t,\ \dfrac{y^2}{b^2} = \cos^2 t$

$\dfrac{x^2}{a^2} + \dfrac{y^2}{b^2} = \sin^2 t + \cos^2 t = 1$

Equation of an ellipse.

47. **(B)** When $x \geq 2$, $y = x - 2 + 2$; or the straight line $y = x$. When $x < 2$, $y = 2 - x + 2 = 4 - x$, which graphs as another straight line. Thus the graph is a pair of rays which form a "V."

48. **(A)** The converse is $q \rightarrow\sim p$. The contrapositive of this is $p \rightarrow\sim q$.

49. **(D)** Conjugate of $1 + i$, namely $1 - i$, must also be a root of the equation.
Thus the roots are -1, $1 + i$, and $1 - i$.
Sum of roots = 1.
Product of roots = $-1(1 + i)(1 - i) = -2$.
Product of roots two at a time
$= -1(1 + i) - 1(1 - i) + (1 + i)(1 - i)$
$= -1 - i - 1 + i + 1 + 1 = 0$,
thus, equation is $x^3 - x^2 + 2 = 0$.

50. **(B)** $\log_r\left(\dfrac{r}{2}\right) = \log_r r - \log_r 2$

$= 1 - \log_r 2$

$\log_r 6 - \log_r 3 = S - T$

$\log_r \dfrac{6}{3} = \log_r 2 = S - T$

Substituting this value in second equation,

$\log_r\left(\dfrac{r}{2}\right) = 1 - (S - T)$

$= 1 - S + T$